From Environmentalism to Transenvironmentalism

From Environmentalism to Transenvironmentalism

The Ethnography of an Urban Protest in Modern Istanbul

Aimilia Voulvouli

PETER LANG
Oxford · Bern · Berlin · Bruxelles · Frankfurt am Main · New York · Wien

Bibliographic information published by Die Deutsche Bibliothek
Die Deutsche Bibliothek lists this publication in the Deutsche Nationalbibliografie; detailed bibliographic data is available on the Internet at <http://dnb.ddb.de>.

A catalogue record for this book is available from The British Library.

Library of Congress Cataloging-in-Publication Data:

Voulvouli, Aimilia, 1975–
From environmentalism to transenvironmentalism : the ethnography of an urban protest in modern Istanbul / Aimilia Voulvouli.
p. cm.
Includes bibliographical references and index.
ISBN 978-3-03911-947-9 (alk. paper)
1. Environmentalism--Turkey--Istanbul. 2. Political ecology--Turkey--Istanbul. 3. Protest movements--Turkey--Istanbul. 4. Istanbul (Turkey)--Environmental conditions. I. Title.
GE199.T87V68 2009
333.72094961'8--dc22
2009017873

ISBN 978-3-03911-947-9

Cover image: A banner hanging in one of Arnavutköy's main streets, stating 'No, to the interests underlying the bridge!' (photograph by A. Voulvouli).

Hochfeldstrasse 32, CH-3012 Bern, Switzerland
info@peterlang.com, www.peterlang.com, www.peterlang.net

Printed in Germany

To Stratos

Never doubt that a group of thoughtful, committed citizens can change the world. Indeed, it is the only thing that ever has.

— Margaret Mead

Contents

List of Figures ix
Acknowledgements xi
Transliteration xv
Abbreviations xvii

CHAPTER 1
Introduction 1

CHAPTER 2
Turkey and Istanbul: History, Politics and the Struggles of a Global City 31

CHAPTER 3
Arnavutköy District Initiative: Environmentalism, Transenvironmentalism and the Identities of an Urban Protest 55

CHAPTER 4
The ASG in the Turkish Transenvironmental Campaigns Context 97

CHAPTER 5
ASG: A Social Movement Organisation between 'Civil Society' and the State 119

CHAPTER 6
Conclusions: From Environmentalism to Transenvironmentalism 163

References 171
Index 189

List of Figures

The figures are located between pages 96 and 97.

Figure 1	One of Arnavutköy's narrow streets
Figure 2	Arnavutköy after a snowfall
Figure 3	The Bosphorus Bridge viewed from Arnavutköy
Figure 4	ASG participants during the Marmaray Inauguration holding banners and placards
Figure 5	A wall of one of Arnavutköy's timber houses with the statement: 'The Third Bridge will destroy our Arnavutköy'
Figure 6	ASG participants during the Marmaray Inauguration Ceremony
Figure 7	Two banners hanging above Beyaz Gül Street in the heart of Arnavutköy

Acknowledgements

Because this book started life as a PhD thesis I would first like to thank my supervisors, Dr Allen Abramson, Dr Charles Stewart and Professor Chris Pinney, for their guidance and support. Dr Deanna Trakas has always been there for me, not only as my supervisor to my undergraduate dissertation, but even after that. This is why I want to extent special thanks to her for everything she has done for me, and especially for her input in the present work. Dr Barbara Cellarius, Dr Dimitrios Theodossopoulos and Professor Maria Kousis have been very helpful by sending me their works as soon as I wrote to them, in spite of their pressing engagements. I also would like to thank Dr Michalis Psimitis and Dr Manussos Marangudakis for providing me with valuable bibliographical support, among which their own writings feature prominently.

I am grateful to the UCL Graduate School for funding my Turkish courses at the Language School at SOAS, as well as to the Greek State Scholarship Foundation (IKY) for funding my PhD studies and Dr Eleni Papagaroufali, who was the Foundation's appointed supervisor. I would also like to thank the staff of the Greek Embassy in Ankara for their suggestions and help regarding the acquisition of a research visa. Sabancı University and Dr Leyla Neyzi were kind enough to grant me an affiliation at a critical time for the future of my research.

My first months in Turkey were less lonely due to the company of my very kind friends Mustafa, Ulaş, Hüseyin and Mustafa. I am thankful for the company of Dr Gabriel Ignatow and his wife Neva; I will always be grateful to them for making my days more colourful and more Turkish. Special thanks go to Dr Ignatow for his valuable intellectual contribution to my work.

At this point I would like to thank everyone who has been kind enough to sacrifice their time and give me an interview regarding the Third Bridge issue, including the Turkish Ministry of Public Works, the Press Representative of the Municipality of Beşiktaş and the members of the Istanbul

Chamber of Architects, who not only answered my questions but gave me access to their archive.

But this book would not have been written if it were not for the people of Arnavutköy whom I admire and respect, not only because they have the courage to stand up for what they believe in but also because they have dedicated much of their precious time – for their protest – to introduce me to the issue. However, one thing for which I will be forever grateful to them is the fact that they honoured me with their friendship, and this is why no thanks will ever do justice to the contribution of the following persons: Özden Hanım, Fezal Hanım, Ayten, İsmail, Arif, Mahmut Bey, Fulya Hanım, Mürsel Bey, Binnur, Nejla, Almıla and Tarik. Thanks to them, I had the chance to see Istanbul with different eyes and understand why in my mother tongue, Greek, it is the only city (*Poli*) referred to with a capital 'C' ('P' in Greek). At this point I would like to express my gratitude to a person who 'left' us but I am sure will always be remembered. Professor Günhan Danişman has been a guiding light for me during my fieldwork. Even though it has been years since I lived in Istanbul, the grief for his loss is felt as though I was right there right now.

I also thank Professors Haluk Gerçek, Nadir Yayla, Ergun Geldizlioğlu and Çağlar Keyder for their scientific insights on the subject, as well as Dr Katerina Nikolarea for her valuable corrections on the first and last draft of this book. The people of Peter Lang Publishing, Nick Reynolds, Kara Dolan and Shirley Walker Werrett have done a great job in helping to turn the initial manuscript into a book and for that I am grateful to them. The scientific committee of the International Conference on Nature, Science and Social Movements held by the Department of Sociology of the University of the Aegean in Greece has given me the opportunity to present my first findings immediately after I completed my fieldwork, and for that I wish to thank them all. In addition, I am grateful to the participants of the workshop on Environmental and Ecological Issues in Cities at the 8th EASA Conference, held at the University of Vienna in September 2004, for the constructive discussion that followed my presentation. Special thanks go to the Department of Social Anthropology and History at the University of the Aegean in Greece, where I completed my undergraduate studies and which I consider my intellectual 'home' for various reasons.

My studies there have – as we say in Greek – opened up my mind and provided me with a solid anthropological background, which has been my guidance through all these years. Also, I wish to thank the department's MA programme in Social and Historical Anthropology and its co-ordinator, Professor Eythymios Papataxiarchis, for their invitation to present my research during the 4th Annual Meeting of PhD Candidates held by the department in Mytilene on 5–6 July 2007, and also for their constructive criticism.

Last but not least, I thank my beloved family, my mother Eleni Antona, my father Ilias Voulvoulis – who is the person to whom I owe my engagement with social anthropology – my sister, Maria Voulvouli, and my grandmother, Milia. I cannot find words to describe my gratitude not just for their faith in me during this whole task but for their precious and everlasting help, support and love.

This book is dedicated to my husband Stratos Georgoulas. I could not be more grateful that life has brought me to the path of such an intellectual treasure. Through all these long years, his guidance and love have marked not only this book but also, and most importantly, my life.

Following the example of the people of Arnavutköy who write their banners and placards in many languages, to all the above people and institutions I extend a big 'Thank you', *Teşekkürler*, *Ευχαριστώ* ...

Mytilene, February 2009

Transliteration

In the text Turkish words and the names of Turkish authors are written with the use of the Turkish alphabet. Below there are those letters of the Turkish alphabet that have no direct equivalent in the English alphabet, and those which are common to the two alphabets but are pronounced differently (Rona, B. 1999):

ı is pronounced roughly like the –er combination in some English words

ö is pronounced like the vowel in the English word 'bird' but shorter

ü is pronounced as the vowel –i but with rounded lips

c is pronounced like the English –j

ç is pronounced like the English –ch

ğ is called soft g (*yumuşak g*); it serves to lengthen the vowel preceding it

ş is pronounced like the English –sh

Abbreviations

AKP	*Adalet ve Kalkınma Partisi* (Justice and Development Party)
ANAP	*Anavatan Partisi* (Motherland Party)
APD	Accession Partnership Document
ARIT	American Research Institute in Turkey
ASG	*Arnavutköy Semt Girişimi* (Arnavutköy District Initiative)
BAD	*Boğaziçi Arnvutköylluler Derneği* (Bosphorus Association of Arnavutköy)
BSNN	Black Sea NGO Network
CHP	*Cumhuriyet Halk Partisi* (Republican People's Party)
CSO	Civil Society Organisation
DHKD	*Doğal Hayatı ve Koruma Derneği* (Association for the Protection of the Natural World)
DP	*Demokrat Parti* (Democratic Party)
EAA	Economic Assistant Act
EFT	Environmental Foundation of Turkey
EIB	European Investment Bank
EIAR	Environmental Impact Assessment Report
EU	European Union
FoE	Friends of the Earth
GSMT	Global Social Movements Theories
ICOMOS	International Council on Monuments and Sites
IPPNW	International Physicians for the Prevention of Nuclear War

İTÜ	*İnstanbul Teknik Üniversitesi* (Istanbul Technical University)
KHRP	Kurdish Human Rights Project
LSE CCS	London School of Economics Centre for Civil Society
LoC	Library of Congress
MDP	*Milliyetçi Demokrasi Partisi* (Nationalist Democratic Party)
MHP	*Milliyetçi Hareket Partisi* (Nationalist Action Party)
MÜSİAD	*Müstakil Sanayici ve İşadamları Derneği* (Independent Industrialists' and Businessmen's Association)
NGO	Non-Governmental Organisation
NIABY	Not In Anybody's Back Yard
NIMBY	Not In My Back Yard
NSMs	New Social Movements
RMT	Resource Mobilisation Theory
SMO	Social Movement Organisation
TEMA	*Türkiye Erozyonla Mücadele Ağaçlandırma ve Doğal Varlıkları Koruma Vakfi* (Turkish Foundation for Combating Soil Erosion, for Reforestation and the Protection of Natural Habitats)
TURSTAT	Turkish Statistical Institute
TÜSIAD	*Türk Sanayicileri ve iş Adamları Derneği* (Turkish Industrialists' and Businessmen's Association)
UNEP	United Nations Environment Programme
UNESCO	United Nations Educational Scientific and Cultural Organisation
USA	United States of America
YTL	*Yeni Türk Lirası* (New Turkish Lira)
WSSD	World Summit on Sustainable Development
WWF	World Wildlife Fund

CHAPTER 1

Introduction

Collective actions against the state or private interest companies have become an increasingly common phenomenon in both developed and developing countries. From local grass-roots to national and transnational campaigns, such actions problematise the concepts of politics and democracy in terms of not only who has the right to decide, but about what and for whom. Decisions to initiate large-scale projects such as hydroelectric dams, highways, mines and power plants are often met with opposition from the members of the community in which these projects are planned. Collectivities of individuals publicising their concerns about the effects of the projects on the health, social life and economics serve to motivate the formation of opposition campaigns. Social scientists from a variety of disciplines (social anthropology, sociology, political sciences, geography, urban studies, economics) who have studied these collective actions have contributed to the understanding of social mobilisations and their influence on social change. Examples of research about major civil protests against the construction of hydroelectric dams include studies of the Movimento dos Antigidos por Barragens (MAB) in Brazil (Cummings 1990; McCormick 2006; Spyridaki 2007), the Narmada Bachao Andolan (NBA) in India (Arundhati 2000; Spyridaki 2007), and organisations in Turkey (KHRP 1999, 2000; Ingatow 2005, 2006). Other mass actions include the M11 link road protest in East London, UK (McKay 1996; Derek 1999), the Hindmarsh Island Bridge Campaign in Australia (Kenny 1998), and the Bergama mobilisation in Turkey against the operation of a goldmine (Sachs 1997; Öncü and Koçan 2001, 2002; Arsel 2003; Özdemir 2003; Beşpınar-Ekici and Gökalp 2006).

This book presents an anthropological study of a grass-roots protest group in a neighbourhood of Istanbul, Turkey. The mobilisation began following the announcement of plans for the construction of a third bridge over the Bosphorus Strait which would connect the Asian with the European shores of Istanbul. In opposition to the construction of the bridge, the residents of the European neighbourhood (Arnavutköy) in which foundations of the bridge would be placed organised an initiative called Arnavutköy District Initiative – in Turkish, ASG (*Arnavutköy Semt Girişimi*). According to the participants of ASG, the reasons for their resistance concerned the destructive effects that the construction of the bridge would have on the area's natural and cultural assets as well as on the life of its residents. The neighbourhood provided an ideal setting for my ethnographic fieldwork in order to explore the ways in which anthropological perspectives on local protest actions and organisations can contribute to an understanding of social movements.

An anthropological examination of mobilisations such as the ASG has the advantage of reminding us that every similarity hides more than one difference (Appadurai 1996: 11). Social movements are not homogeneous collectivities; they are rather what Arjun Appadurai would describe as 'neighbourhoods'; that is, 'social forms in which locality as a dimension – is constituted by a series of links between the sense of social immediacy, the technologies of interactivity and the relativity of contexts – is invariably realized' (ibid.: 178). Nevertheless, 'however deeply a description is embedded in the particularities of place, soil, and ritual technique, it invariably contains or implies a theory of context – a theory, in other words, of what a neighbourhood is produced from, against, in spite of, and in relation [to]' (ibid.: 184). In cases like the one this book focuses on, the collective action of the neighbourhood of Arnavutköy can be seen as a result of changes taking place from the effects of economic and cultural globalisation[1] (Psimitis 2006). These effects can be seen in any number of social movements organised around, for example, human rights, feminism, consumers' rights,

1 According to Habermas, in an undoubtedly globalised world – economically and subsequently culturally – there is an increasing awareness of capitalist penetration in areas of life, traditionally protected by it and detached from the values of capitalist society (in Psimitis 2006).

ethnic–religious–cultural minority rights, sexual emancipation, community participation, urban action, and environmental issues.

On methodology

The first contact

Although this book eventually evolved to be about an urban 'transenvironmental' (Kousis and Eder 2001) protest, with only a few references to bureaucracy, in my first year as a PhD candidate my intention was to investigate the perceptions of bureaucrats about the environment compared to those of other citizens. Therefore, I began searching the Internet to identify cases of disputes between the Turkish state and its citizens which led to the discovery of the 'Third Bridge' conflict. Initially, I reviewed this conflict to determine if and how the Turkish government engaged in environmental discourses. The first thing I read was that Arnavutköy, according to a government decree, was one of the sites designated for special protection (SIT area), as is the whole Bosphorus Strait. Bearing this in mind, I was taken by surprise when I found out that in that specially protected area the Turkish Ministry of Public Works and Settlement (*Bayındırlık ve İskan Bakanlığı*) was planning to construct a concrete and steel suspension bridge. My previous impression was that a protected status does not allow any intervention that would dramatically change the character of an area, and, in my mind, a huge suspension bridge would definitely to do exactly that.

My subsequent research discovered that this was not only my impression, but apparently a complaint of the residents of the area, supported by the fact that they had organised a protest to address the issue. I decided to visit the area and see if it would be suitable for my fieldwork. I travelled by bus from Taksim Square with an American friend and we went directly to the American Research Institute in Turkey (ARIT) – which was coincidentally based in Arnavutköy – to ask whether there was information about the issue. The director was well informed and he gave me some telephone numbers of people actively involved with the anti-bridge campaign. Before contacting them, we strolled in the neighbourhood where we saw banners hanging from rooftops declaring the residents' opposition to the construction of the bridge ('No to the interests underlying the bridge!' (*Çıkar köprüsüne Hayır*) '3rd Bridge: Hayır, No, Nein, Non' (*3. Köprü: Hayır, No, Nein, Non*); see Figure 6). I contacted one of the participants[2] of the initiative who later became one of my main informants and I told him that I was interested in their protest as well as that I was seriously thinking of focusing my PhD research on the matter. He was very eager to talk to me, as was almost everyone I approached during my fieldwork, and we scheduled an appointment a few days later. After the appointment ended, I was convinced that Arnavutköy would be the site of my fieldwork.

Doing fieldwork in a city

In 1975, Jack Rollwagen wrote that in order for the study to become significant anthropologists must place their investigations 'of one social form, of one neighbourhood, of one city, and/or of one region within a nation, into the context of the nation-state or a region larger than the nation-state' (Rollwagen 1975: 4). I find his comment quite relevant to my study as I believe that the Arnavutköy initiative should not be seen as isolated from larger geographical, historical and political contexts, but as part of what Kemper calls 'international urban systems through time and space' (1991b: 374). After all, Arnavutköy is part of one the largest cities on the planet, Istanbul; according to the 2000 census, the main city's population is listed at 8,803,468 inhabitants, and 10,018,735 if the peripheral provincial areas are included (http://en.wikipedia.org). In addition, as Moore (1996) argues, there are certain organising principles shared by all cities which create

2 In order to protect the privacy of my informants, any name used in the text is not the actual name of the person discussed in each section.

an urban network of distinctive social–cultural and political–economic domains (Kemper 1991b: 374).

Keeping in mind the similarities which Istanbul and Arnavutköy, as part of it, share with other large metropolitan centres, in socio-cultural as well as political and economic domains, my methodological approaches during fieldwork followed a pattern wherein ethnography moved from its conventional single-site location contextualised by macro-constructions of a larger social order (such as the capitalist world system) to multiple sites of observation and participation (Marcus 1995: 95). Even though my fieldwork was centred on Istanbul, I also saw it as a global city (as it had always been in my mind), where the dynamics and processes which became territorialised were also global (Sassen 2001: xix). This means that ASG should not be seen as a unique case of a protest action but as one of many protests occurring around the globe in cities with similar socio-economic and demographic development.

World System Theory has extensively analysed the global character of economy from which European capitalism resulted, beginning in the fifteenth century. As theorists of the World System School (Wallerstein 1974) and those influenced by it (Featherstone 1990) claim, concentrations of capital are centred in cities and, in fact, create cities (Kutsche 1989). In this sense, the ASG protest should be seen as a protest against a certain kind of capitalist development related to processes of globalisation. This perspective has two advantages. First, it allows for the examination of a development project (such as a bridge in a developing nation) as a construction produced by transnational corporations. This is applicable to the present case, since the construction of the Bosphorus bridges was assigned to foreign companies. Secondly, it privileges an analysis which attempts to depict the full complexity of social life in cities (Sanjek 2004). In this sense, it enables multisited research (Marcus 1995) since it inevitably considers the anthropologist's main subjects of study – the people and, in this case, the ASG participants – as involved in a multilevel discourse produced in several different locales (local and global).

Following Marcus' techniques of multisited ethnography, my research revolved around various aspects of the same issue. My fieldwork in Turkey lasted eighteen months, eleven of which I spent in Arnavutköy. As mentioned above, it began as a preliminary investigation mainly through the Internet. After the first meeting in Arnavutköy, I settled in the area, initially in the ARIT guesthouse and later on, in the house of one of my informants. I continued my research after I had left Istanbul by keeping in contact with my informants, receiving newsletters by ASG and keeping up emerging events related to the issue of the bridge through the electronic press. While in the field, the main methods of my data production were multilevelled, including participant observation, interviewing, collecting news articles, travelling within the country, and keeping the classic ethnographic diary.

It would be misleading to claim that my fieldwork was directed by a detailed research plan. Apart from travelling in Turkey to become familiar with the country and settling in Arnavutköy, all the other parts of the field research emerged during my residence in Turkey. Soon after my arrival, I came to realise that conducting ethnographic research on the 'particular discourse of policy requires different practices and opportunities than to do just fieldwork among the situated communities such a policy affects' (Marcus 1995: 100). Since my focus was the conflict between the Turkish government and the residents of Arnavutköy, I needed to take into account the point of view of both sides. In order to obtain this kind of information, I needed to use totally different research techniques for each side. While I would not claim that my book constitutes an ethnography of Turkish bureaucracy, I did devote a large part of my time to interviewing bureaucrats from a variety of locations. They were from Arnavutköy and the surrounding administrative districts (*muhtarlıks*); the Municipality of Beşiktaş (*Beşiktaş Belediyesi*) to which Arnavutköy administratively belongs; and the Ministry of Public Works and Settlements. I kept in mind that a study of bureaucracy constitutes a different research domain; thus, my investigation followed a different pattern and looked for different signs.

The difference between interviewing bureaucrats and participants of ASG or residents of Arnavutköy is illuminated when it is realised that for the latter, interviews with journalists (some of them residents of the neighbourhood) had become virtually part of their daily culture; in fact, giving interviews had been one of their main opposition strategies. Consequently, 'interviews' about the opposition to the third bridge were actually participant observation. Even though interviewing is usually seen as a secondary technique to produce data to supplement participant observation which is the primary method, in this case I used interviewing as a way of participating in the neighbourhood's life. Hence, participant observation was also multi-sited. It concerned both participation in everyday life through my housemate, participation to ASG's weekly meetings and its other activities (dinners, demonstrations, discussions) and also interviewing them. Apart from these activities, I also had the chance to interview visitors of the area, scientists, NGO members, representatives from the Chamber of Architects, artists and activists.

George Marcus's multisited ethnography consists of techniques which he entitles 'following'. In my research I engaged in activities that he designates as 'follow the metaphor'. His suggestions include observing the circulation of signs, symbols, and metaphors relating to the subject of study. Therefore, aside from the interviews, I followed the Third Bridge issue as it appeared in the popular press. Through the archives of the ASG, the Istanbul Chamber of Architects (*İstanbul Mimarlar Odası*), the electronic records of national and international press as well as the hard-copy national press, I collected articles referring to the Third Bridge issue. My aim was to identify the verbal practices and the rhetoric used to speak about the issue. In Marcus's words, I tried to 'trace the social correlates and groundings of associations that are most clearly alive in language use and print or visual media' (1995: 108). Keeping a diary while I was living in Arnavutköy was not only for writing down things to remember for future reference but also to incorporate my informants' biographical data in a more coherent way than the interview text. This technique helped me create an ethnographic space in which the issue of the conflict over the construction of the third bridge was seen – as possible as this can be – through the eyes of the people opposing the bridge. Finally, travelling in the country made clear for me the existing differences between Istanbul and the rest of Turkey and helped me see the larger picture to which Istanbul and Arnavutköy belong.

The anthropologist and the field

As a Greek citizen, choosing Turkey as the country where my fieldwork would be based held personal significance. My paternal grandfather was born in Dikilli, a town on the Aegean coast of Turkey. His father, who was professionally active both in Asia Minor and on the island of Lesbos, decided to move to the island which at that time was part of the Ottoman Empire. My family's migration to Greece (before Lesbos became part of it) was not, as in many other cases, a violent one; hence our memories are not bitter towards the opposite coast or *karşı* (even today, many inhabitants of Lesbos use the Turkish term *karşı* for 'opposite' in general, not only to refer to the coast of Turkey.) When I was a little girl I discovered that our family name was in fact of Turkish origin, with a slight Hellenised touch: Voulvouli is derived from *Bülbül*, which is a common surname in Turkey and literally means 'nightingale'. From my part, there had always been the curiousity to visit the *karşı*. From the few times I had been to Turkey, I realised that visiting as a tourist was not enough to satisfy my curiousity and I decided that at some point I would go there to live for a while and learn Turkish. It was my belief that by doing that I might be able to discover my roots, an idea which faded while I was an undergraduate student in the Department of Social Anthropology at the University of the Aegean. Even though my studies in anthropology directed my interest in the country to more 'scientific' pathways, it was during that time when I began to envision a way to stay in Turkey for more than a few days. I would go as an ethnographer.

Given the historically hostile climate between the two countries, I was aware that there were presuppositions on both sides regarding the other. Even though I grew up in a family where nationality was never the most important symbol of identity, I was a little afraid that being Greek could be a hindrance to my research. Nevertheless, I considered it as a challenge to discover how I would interact with Turkish people on a long-term basis and how they would react to my presence as well. My fears disappeared after the first weeks of my residence in the country. What I faced in Turkey was simply what any other researcher or foreigner faces. In fact, I often felt that many Turkish people saw Greece as an example to be followed (especially as for its integration in Europe) and that the Greeks were the 'good' neighbours of the West as opposed to the 'bad' neighbours of the East.[3] However, the issue of why I did not wear a veil gave me a contradictory message.

Occasionally, when in the company of women in a private home, I was asked if I veiled my head when in public. 'Are you covered?' (*kapalı*) I was asked sometimes. They were surprised when I explained that as far as I knew women in Greece cover their heads for a number of aesthetic and practical reasons but it is mainly the elderly women (and rarely) who cover their heads for religious purposes and mostly in church. One woman who did not wear a veil commented to me, 'I thought that Greek women were like us' but the context of 'us' referred to the veiled Muslim women of Turkey, not herself. In other words, she saw Greek women as primarily non-Western as she did Turkish women who wore the veil. On another occasion, while a guest for dinner, I refused to taste a *meze*[4] cooked with wine. The lady of the house (a clearly secularist house), said, 'Oh, I will never understand those religious habits,' and was obviously relieved when I told her that I was actually agnostic, but unfortunately, allergic to wine. The above incidents suggest that there were indeed preconceptions of me. However, those preconceptions were related to my religious preferences rather than my ethnic origins and I must admit that as a person who has always felt a bit different in my own culture, particularly with respect to religious issues (separation of church and state in Greece has never been fully achieved and 95 per cent of Greek population are Orthodox Christians), during my stay in Turkey I often felt annoyed with fanatically secular individuals. It was easier for me to identify with the marginalised believers than the dominant secularised.

3 To that, among other things, I believe that contributed the efforts of the two countries for 'good neighbouring' as politicians call it and the effort of Turkey to join the EU.

4 *Meze* is a selection of appetisers or small dishes taken with alcohol, which can be served either alone or as the first dish of a meal. In Turkey, *meze* is served with *rakı* and usually consists of cheese, spicy aubergines, various salads, *cacık*, meatballs and *dolma*.

My above comments are relevant to what Gefou-Madianou (1998) points out about the fieldwork experience: As a culturally informed subject, the individual ethnographer always carries his/her identities from which it is impossible for him/her to disengage while in the field. Moreover, in my case it was the pre-existing familiarity with the 'other' which triggered my interest in conducting research in Turkey. In addition, even the choice for the content of my research stemmed from my identities and, dare I say, political convictions. Therefore, I agree with those who claim that objectivity in anthropological writing is not possible. On the contrary, in the analysis of my data I tried to be as self-reflexive as possible even if not explicitly, in order to be consistent with the conclusions prompted by my theoretical background; namely that the ethnography of a conflict cannot be seen separately from the historical and political contexts to which both the ethnographer and the informants belong. After all, as Cunningham (1999: 5) claims, 'while anthropologists are in the process of discerning globalisation as an analytical phenomenon, they may also be located in – and therefore subject to – the processes of it'.

This brings into discussion another aspect of ethnography relating to the closeness of ethnographers to their informants. In her discussion of 'anthropology at home', Gefou-Madianou (1998) states that conducting fieldwork in the home culture (or place of the researcher's origins) involves moral and political issues. An ethnic identification between the ethnographer and the informants poses some difficulties which can be translated into a sense of responsibility of the ethnographer towards the informants. Shared opinions between ethnographer and informants, or feelings of gratitude towards informants for opening their homes can also create feelings of responsibility. How could I be objective in writing about the conflict over the bridge when I, silently but nevertheless, supported their struggle? How could I write something less in favour to the protest when the people had opened not only their houses but also their hearts and minds to a foreigner who did not even speak Turkish very well? I soon realised that I could not. As Paré (in Edelman 2001: 26) writes about his fieldwork in rural Mexico:

> For many of us it turned out to be impossible to record acts of repression and forms of exploitation and to witness the difficulties the peasant organisations had in making their voice heard without taking sides [...]. Participation – whether directly in the organisation, in advising groups, in collective analysis with the organisations themselves, in negotiations, in publicity, in solidarity, in communications, or in the government as a planner, functionary or technician – necessarily implies taking a position, a 'committed' vision.

Therefore, I decided to turn to what Marcus describes as a 'circumstantial activist, a condition which results from working in a variety of sites, where the politics and ethics of working in any one reflects on work in the others' (1995: 113). As far as the second ethical dilemma is concerned, that is, the gratitude I felt and still feel for ASG participants, I decided that the only way to feel less guilty for any potential misjudgement of their battle against the construction of the bridge was to do exactly what I stated at the beginning of this methodological account. Any generalisations that follow concern neither ASG nor Arnavutköy as a unique case of protest. This is about the social mobilisation of people defined by multiple dynamics namely cultural, national and socio-political. After all, as Hannerz (1996: 78) reminds us, the multiple, the complex, the ambiguous, the diverse are also socially organised.

On theory

Theories of collective action such as the one developed by Talcot Parsons during the 1920s and the 1930s have examined revolutionary movements as anomic or deviant collectivities (Merton in Psimitis 2006). The symbolic interaction approach of the Chicago School viewed social movements as mechanisms of positive social transformation (Edelman 2001). The Marxist approach linked collective action to the system of production and suggested that it be seen as a protest of the proletariat against the bourgeoisie, while a more individual-oriented approach has been proposed by the Rational Choice School, claiming that the motives of collective action are private.

More recent theories of collective action choose social movements as units of analysis. New Social Movements (NSMs) theories speak about social movements as groups of people unified by common identities – namely, ethnic, gender and professional (e.g. student) – while having shifted their attention from class struggles to environmental and peace campaigns. Practitioners of Resource Mobilisation Theory (RMT) examine how such movements mobilise to achieve their goals.

Within this framework, 'the role of ethnography in the study of social movements has been significant but seldom theorised' (Edelman 2001: 309). According to Edelman, even though ethnographers have produced detailed accounts of collective action, they tend to focus on single-organisation or single-issue collective action rather than focusing on the broader political and social contexts in which such actions emerge. However, not all ethnographic studies of collective action fit this profile and Edelman admits that Latin America ethnographers, such as Arturo Escobar and Sonia Alvarez (1992) and Alvarez et al. (1998), who conducted research among the Latin American social movements (the Zapatistas, grass-roots mobilisations, peasant movements) examined these movements in relation to the state. Furthermore, more recent anthropological studies such as the collection edited by June Nash (2005), focus on social movements as a result of globalisation processes, implicating issues of civil society, religion, de-territorialisation, privatisation, individualisation and global cosmopolitanism. The views of these studies on social movements take into consideration the fact that

> the locus of working-class struggle in the workplace that emerged with industrial capitalism in the nineteenth century is now taking place elsewhere. Shopping mall boycotts, protests against neo-liberal policies of the World Trade Organisation at summit meeting headquarters, and highway blockades by small-scale commodity producers of Third World countries have become symbolic grounds for protest by the dispossessed and impoverished (ibid.: 3).

These studies share similar concerns with ethnographies that do not focus on collective action. Such studies engage with ethnography of the state (Navaro-Yashin 2002) and its relation with religious groups (Shankland

2003), political anthropology (Gledhill 1994; Lewellen 2003) and urban anthropology (Rollwagen 1975; Kemper 1991; Durrenberger 2003).

Within this framework, I intend to discuss the ASG mobilisation as a civil protest against the state. The term 'civil' will not go unquestioned; to the contrary, as I will explain below, the prevailing binary opposition between state and civil society risks oversimplifications not only because Civil Society Organisations (CSOs) use state methods to achieve their goals but also because certain aspects of them confirm that they operate as governmental organisations. Apart from the fact that they are funded by governments, in many cases, they function complementary to them. Therefore, CSOs are agents of politics both within national borders and transnational territories. National identity, essentially linked with the nation-state, gives its place to regional, global and transnational identities which, in turn are, linked with global civil society. In this sense, CSOs assume the character of a systematic collectivity; that is, a social movement which shapes and re-shapes politics and contributes to processes of social change. Therefore, civil society is not as civil as it appears to be. It is linked with politics in terms of 'the processes involved in determining and implementing public goals and in the differential achievement and use of power by the members of the group concerned with these goals' (Swartz et al. in Lewellen 2003: 85). Thus, the demands of social movements can be considered as demands for power in order to fulfil the goals of certain groups of people.

Definitions of a 'social movement' create other predicaments. Those following the NSMs paradigm claim that New Social Movements, as opposed to the old social movements that focused on labour and working-class issues, derive mainly from the symbolic realm of social relationships and demonstrate interactions among collective and individual actors within civil society. Resource Mobilisation Theory (RMT) perceives social movements as rational collective actions using available resources to achieve their objectives. From this perspective, social movements are political agents and their success is measured to the degree they are recognised. More recent theories of social movements, combine elements of both these definitions while rejecting others. RMT focus on viewing collective action as cost-effective calculation in terms of possession of resources, an issue which is of less importance in the set of theories discussed below which examine grass-roots mobilisations, even of the very poor.

Ethnographers, through sustained fieldwork and engagement throughout the area of research (Hann 1995: 160), examine ways people respond to global tendencies in culturally and historically distinct ways (Nash 2005: 13). This effort illustrates that local grass-roots campaigns, which, through the prism of RMT, might be considered interest-oriented and/or NIMBY[5] actions (Psimitis 2006) are a lot more than that. Furthermore, such approaches reject the tendency of NSMs to overlook economic and power inequalities, since they find they play an important role in struggles, especially in non-Western settings (Edelman 2001). Finally, their assumptions stem from research, focusing on how social movements are organised, something that NSM theories – in contrast to RMT – fail to demonstrate (Psimitis 2006). I call this set of theories Global Social Movement Theories (GSMT), since their distinct characteristic is that they focus on the aspect of globalisation in their studies. In other words, they see social movements as tensions created by globalisation processes and, therefore, they manage to combine grand theory with micro-level analysis – a characteristic which means this approach is favoured by social anthropologists.

The case of environmentalism is a credible example of how a social movement can be seen as a result of globalising forces, at both the macro- and the micro-level. Tracing its origins to the nineteenth century, environmentalism has provided one of the most influential values and movements of our age. From the enactment of laws protecting whole areas to the foundation of green parties, the protection of the environment seems to be gaining ground in world politics. Global Summits are organised to discuss the earth's future (Stockholm 1972; Rio 1992; Johannesburg 2002; New York 2005); NGOs are founded in the name of environmental protection; politicians, world-famous artists, businessmen and corporations declare their commitment to the environmental cause and take actions to prove it. Their

5 NIMBY is an acronym standing for Not In My Back Yard that describes the situation in which residents of an area oppose a development because they find it damaging for their place and propose its construction elsewhere (Wolsink 1994).

main argument is that this planet must be preserved for the generations to come. In other words, our actions upon earth must be sustainable.

Contemporary environmentalism advances the notion of 'sustainability' as key condition to the planet's future. Green politicians and businessmen are enthusiasts of this concept. The former President and Vice-President of the United States of America, Bill Clinton and Al Gore respectively, have organised worldwide campaigns to help save the planet from the greenhouse effect and climate changes. Millions of dollars are spent on green investments and thousands of man-hours are spent in global official and unofficial meetings regarding climate change.

Market globalisation seems to be intrinsic to sustainability, and both concepts seem to offer what Agryrou (2005) calls the imperatives of the West serving as a medium to maintain its hegemony over the East; or, paraphrasing that, the imperatives of those in power serving as a medium to maintain their hegemony over the powerless. Therefore in order to understand environmental conflicts we must first keep in mind that many of them are reinforced by citizens whose rights are overlooked and, secondly, that the groups of citizens organised in such cases have a transnational character because the forces they are opposing to are transnational, too. In this context, the increasing importance of supranational powers, such as the European Union, overcomes the influence of nation-state. Examples in the Turkish context are religious, Kurdish and environmental movements (Şimşek 2004).

The Turkish context

Manuel Castells (1983: 42) claims that 'it is impossible to understand the meaning of a movement without referring to historical conditions in which the movement appears'. Following this, in this book I introduce Turkey as a historical socio-political context in which ASG was born. Turkey became a nation-state in 1923 and ever since the changes at institutional, economic and cultural level have been radical and continuous. Until the 1938 death of the founder of the Turkish state, Mustafa Kemal Atatürk – who declared that Turkey belonged to the West – the country has experienced both political and cultural changes, reflected in institutions, language, attire, gender relations and religious habits.

Since 1938 Turkey has faced the transition to a multiparty political system and a sequence of political unrest, repetitive elections and *coups d'état*. At the same time there has been the economic opening of the country to the West, a critical moment of which was the Marshall Aid received in 1948. With an obligation to be accountable to the donors, Turkey followed a certain kind of development, which through the years resulted in rapid economic and demographic change, urbanisation being the most apparent characteristic. The tradition initiated by the State Highways Department, created during the 1950s under US supervision (Mango 2004), was followed by all the subsequent governments (see Keyder 1999). More specifically, the Özal governments made their mark in history as the administrations which invested large sums of money in the construction of highways and new technologies (Şimşek 2004).[6] Turkey's integration into global processes followed these changes.

The global character of Turkey becomes more conspicuous in big cities such as Istanbul, which are characterised by their efforts to attract foreign investments. In order to achieve this such cities prioritise developments that aid their efforts. The construction of business districts and the roads which facilitate access to them form examples of this kind of development. In the case of Istanbul, the construction of the existing Bosphorus crossings (the Bosphorus Bridge and the Fatih Mehmet Sultan Bridge) can be seen as such examples. The business districts of Istanbul (i.e. Levent and Takism) are situated in the European side of the city and the Bosphorus bridges along with the adjacent roads facilitate the access of those who live in Asian Istanbul (which is mainly a residential area) to their workplaces.

However, such developments resulted in the displacement of many people and this eventually created conflicts such as the one on which this

6 Many of my informants referred to ex-Prime Minister Özal as 'the king of the bridges'.

book focuses. Therefore, in the pages that follow, I examine the ASG anti-bridge campaign not only as part of the so called 'global civil society' but also as a result of the global character of Istanbul.

Environmental associations and transenvironmental campaigns

Since the aftermath of the 1980 *coup d'état*, Turkey has been facing a growth of the so called civil society. A series of Islamic organisations, business associations, Kurdish, Alevi, environmental and feminist groups have mushroomed (Kubicek 2005; Ignatow 2005a, 2005b). The interpretations of this phenomenon vary. Some claim that the development of civil society in Turkey was brought about due to the emergence of new actors, new mentalities and the new language of modernisation. Others consider it a result of the EU harmonisation process and the 'emergence of alternative meanings of modernity' (Keyman and Içduygu 2003). Finally, those less optimistic claim that all these new groups function as a substitute to the void left from the repression of associations organising around leftist politics by the military regime of the 1980s (Beşpınar-Ekici and Gökalp 2006).

All the above are plausible explanations for the development of Turkish civil society. To these, I would add my view of CSOs in Turkey, following Navaro-Yashin (2002), who claims that the presence of political in Turkish 'public life' is too strong to be ignored by students of civil society. For her, talking about civil society contrasting it to the state risks erroneous conclusions. This is why she replaces the notion of civil society with the notion of 'public life', which refers 'both to people and the state, by producing and recasting the political, thereby enabling an analysis on both people and the state as part of the same domain' (ibid.: 2). Following this, I see Turkish CSOs as results of their interaction with the state. The 3,000 existing CSOs in the country either oppose to or function supplementary to state policies. Nevertheless, either as challengers or as facilitators of state policies, they both have the latter as their main point of reference.

This characteristic of Turkish CSOs has its roots to the distinct political context of Turkey, which changed drastically after the 1980 *coup d'état*, during which half of the existing NGOs were closed down (Şımşek 2004) and the remaining were depoliticised (Beşpınar-Ekici and Gökalp 2006). It was as late as 2004 that a more moderate Law on Associations came to replace the dogmatic prohibition of linkages between NGOs and politics. According to this law, state intervention on civil society activities is now regulated and the collaboration between NGOs and political organisations is allowed. These developments were perceived by a part of Turkish society as advancements, as far as democratisation, modernisation, liberalisation and opening of the country are concerned. This interpretation increased the popularity of CSOs, a fact strengthened by their successful mobilisation during the 1999 fatal earthquake that hit the Marmara region of the country.

However, there are those who claim that civil society in Turkey exists 'more as a slogan than as a reality', since it is still controlled by the state and the individual participation in decision-making is limited (Kubicek 2005), a fact which results in a rigid state–society polarisation (Özdemir 2003). Furthermore, even though the new Law on Associations is more liberal, the vast majority of CSOs are subject to the earlier Law of Associations. This situation is different in the cases of CSOs that focus on EU implementation projects and other international funding sources. In addition, economic, political and cultural globalisation has created new paths of civic actions which are reinforced by the state, too. An example is the 2001 National Programme, which was created by the government to facilitate the implementation of the Copenhagen criteria set by the EU as condition for the country's accession to the European Union.

In this context, it has been acknowledged that there are issues that cannot be solved by political parties alone and that in Turkey, as much as in other countries, actors such as civil-society organisations should behave as 'actors of politics', something that is a real phenomenon. CSOs in Turkey are involved with big societal issues, such as Kemalism, modernisation, the protection of the secular and democratic Turkey or Islamic order, Islamic life and other issues prominent in state agendas (Keyman and Içduygu 2003). In this way, CSOs function, on the one hand, as facilitators and in some cases legitimators of state policies and, on the other hand, supplementary to the state, especially in the aftermath of the devastating impacts, as far as the welfare state is concerned, of neo-liberal politics followed since the 1980s (Beşpınar-Ekici and Gökalp 2006). Therefore, CSOs in Turkey should be examined not only in relation with or in opposition to but also as a substitute for the state (Jalali 2002; Beşpınar-Ekici and Gökalp 2006).

As mentioned later in Chapter 4, examples of the Turkish context, such as project implementations from international funds and religious prohibitions, illustrate the articulation between the state and civil society. For example, despite the fact that religion in Turkey is left completely outside the state, religious issues are highly politicised – even more than in countries that have no state–religion separation. Today, in Turkey a large majority of Turkish citizens seem concerned with religious behaviour such as religious ceremonies, religious attire, and the extent to which these affect the secular character of the Turkish state. Moreover, those who are against religious expression count on the army to ensure that religion would not affect the secular character of the state. The paradox of this phenomenon is that political parties (state) and the people (civil society) proclaim the same beliefs – the statist secular beliefs which a state institution itself, the army, is in charge of protecting. This example illustrates that the consideration of state and civil society in Turkey as two distinct realms can lead to erroneous conclusions. It is only when we examine them as the same realm and focus not on their separation but on their interaction that a clear picture of non-state organisations and social movements will emerge.

The case of environmental CSOs in Turkey, some of which act as supplements to the governmental administration and the rest protest against the state is an example of the above claim. Environmental organisations in Turkey began to appear after 1909 when the Law of Associations (*Cemiyetler Kanunu*) was issued. Until the 1930s a number of associations was founded under the supervision of the ruling and only party (CHP), a fact that did not allow them to act as pressure groups. That situation was slightly altered with the beginning of the multiparty period (Özdemir 2003).

However, as Turkey's environmental problems began to grow, governmental environmental institutions also started to be established. The Under-Secretariat of Environment was formed in 1978 and in 1991 the

Turkish Ministry of Environment and Forestry (*Çevre ve Orman Bakanlığı*) obtained portfolio status. Today, Turkey faces problems concerning the urban environment, such as air quality, water supply and water waste as well as solid waste management (Ignatow 2005a); natural resources management, such as water resources, soils and land, forests, biodiversity, the marine and coastal resources, cultural and natural heritage; and natural as well as man-made environmental hazards and noise pollution (Özdemir 2003).

As might be expected, Turkish environmental organisations could have focused on these issues, but this is not usually the case. Except for the fact that Turkish environmentalism is influenced by Western environmentalism, meaning that the perception of the present environmental problems is often infiltrated with Western values, other parameters should be taken under consideration in the study of environmental awareness in Turkey. These are the articulations of environmental demands with claims of democracy and human rights, the Muslim view of environmentalism, as well as the different versions of environmentalism according to different political persuasions (Özdemir 2003).

In light of the above, Turkish environmentalism should be examined as a multifaceted body constituted by industrialists, intellectuals, activists, religious groups and political active individuals and groups. As in other parts of the world, Turkish environmental organisations are not homogeneous. According to Özdemir (2003), they can be divided in three distinct groups: official sector organisations (e.g. Ministry of the Environment), civil environmental organisations (e.g. foundations, associations and cooperatives formed by the private sector) and the political environmental organisations (e.g. the Green Party). The Green Party was founded in 1988 but was banned in 1994. A division of it formed the SOS Mediterranean Association in 1999, whose claims relate environmentalism to democratisation as a means to achieve harmony between man and nature. Another association with similar concerns is the Turkish Foundation for Combating Soil Erosion for Reforestation and the Protection of Natural Habitats (TEMA). This association was established in 1992 by the Turkish businessman Hayrettin Karaca and associates environmental degradation with social degradation. Furthermore, it stands critically to modernist development and consumption patterns. Its modus operandi involves lobbying governments, political parties and education institutions as well as state and private organisations. Another officially constituted environmental organisation is the Environmental Foundation of Turkey (EFT), established in 1978, which aims to increase environmental awareness through research. EFT is linked with the UN, the EU the World Bank and the Organisation of Security and Cooperation in Europe.

Another group of environmentalists in Turkey finds expression in religious environmentalism. According to the Islamic view, environmental protection can be achieved through everyday practice since environmental degradation is the result of modernity. However, Muslim environmentalists, in contrast to political Islam, are not organised in any form of civil association due to the state's suspicion of any religious group except for Alevi environmental associations, which offer a combination of religious and scientific environmentalism. Such groups organise environmentally related competitions and tree-planting campaigns and their members are involved with grass-roots environmental campaigns.

Nevertheless, the above associations are not identified with the grass-roots campaigns. Environmental associations in Turkey, are institutionalised groups with state alliances as well as links to industrialists, the UN, the EU and with international NGOs. On the other end of this model, there are local environmental campaigns opposing to particular issues with minor political or economic alliances but with strategies that achieve the mobilisation of big parts of the society. Such campaigns are the much studied Bergama protest, the Beyoğlu initiative, the İlisu Dam and the Tunceli Dam opposition campaigns, the Akkuyu resistance and the Arnavutköy anti-bridge struggle.

Transenvironmental campaigns

The Bergama protest concerns a conflict between the residents of the Bergama area near Izmir and the Normandy Mining Corporation. The conflict centres on the Eurogold Project for the establishment of a goldmine whose method of gold extraction involves cyanide usage, an element dangerous for public health. The opposing residents reacted in various ways and they even managed to get the case to the Turkish Constitutional Court, which revoked the company's licence. The Bergama case has been examined as a case of multilayered governance, as a human-rights protest and as a case of environmental activism.

Another case is the conflict between the Metropolitan Municipality of Istanbul and the Chamber of Architects over the so-called Tarlabaşı Demolitions. The municipality, as it finally transpired, had planned the transformation of one of the main streets of the Beyoğlu district of the city, İstiklal Caddesi (Independence Street), into a pedestrian street and the opening of a parallel artery which would ease traffic congestion. For this construction to occur, a large number of nineteenth-century buildings would have to be demolished. The conflict did not only concern the clash between conservationists and modernisers, but the clash between the military regime of the 1980s and those who criticised it (Bartu 1999).

Similar issues form the basis of other conflicts such as the İlisu Dam case; this concerns the governmental decision to construct a dam that would flood the city of İlisu, which is of historical and archaeological significance, and that would prompt the relocation of a great number of people. According to those who opposed the building of the dam, the results of its construction were threatening to the local cultural heritage. Like the Beyoğlu case, the İlisu Dam opposition involved more issues than environmental and cultural heritage protection as the area, populated mainly by Kurdish people, is a symbol of Kurdish culture and resistance against the Turkish state. The outcome of this conflict was the cancellation of the construction of the dam, something which was not achieved by the Tunceli anti-dam activists in a similar case of anti-dam protest in a mainly Kurdish-populated area. The Tunceli area is known for its socialist and anti-government groups, a fact suggesting that the opposition to the dam was part of a wider, in terms of issues, opposition. Similarly, the Arnavutköy anti-bridge campaign which is the focus of this book is a conflict involving much more than the objection to the bridge construction.

The Arnavutköy anti-bridge campaign: An urban/transenvironmental social movement organisation

Arnavutköy: The Albanian village

Arnavutköy was inhabited by Greeks during the fourth century; at that time is was known as Estia or Mihailion (www.megarevma.net). After Istanbul passed to Ottoman rule, Albanian (*Arnavut*) families were made to settle in the area and hence the name of the village (Arnavutköy in English translates as 'Albanian village') (Keyder 1999). The site was formerly called Mega Revma (Greek for 'strong current') undoubtedly related to the strong currents in the Bosphorus which result from the flow of water from the Black Sea to the Marmara Sea near the shore of the village. Arnavutköy used to be famous for its strawberry crops and its timber houses. The ones built by the sea,[7] the *Yalıs*, were used as summer houses, but many were burned down in a big fire in 1887 (www.megarevma.net).

Arnavutköy was also and still is famous for its churches. The Profitis Ilias and Taxiarchis churches were built when the Greek community was still large. There is also a mosque, built by Sultan Mahmut in 1832. It is a small single-minaret mosque attached to the police station and close to the oldest fountain of the area. A bath, which was ruined in 1930 (www.megarevma.net), was part of the village's social life.

During the seventeenth century, the area was populated by Jewish people, but by the end of the eighteenth century almost the entire population of Arnavutköy was Greek-speaking. After the eighteenth century a small population of Armenians came to the area. Just before the First World War there were recorded 342 Armenians and 5,973 Greeks living in Arnavutköy (ibid.). Many Greeks left the country between 1914 and 1924, but the big flow out of the area and Istanbul came during the 1950s and the 1960s. During that period many so-called Rum Orthodox people,

7 That edge does not exist anymore due to the construction of an expressway (*kazıklı yol*) in the mid-1980s.

who were excluded from the exchanges agreed in the Lausanne Treaty,[8] moved to Greece mainly due to the hostile climate prevailing the two countries, which escalated on 6–7 September 1955 (Septemvriana) (Christides 2000). In the 1960s the Turkish government expelled many Greeks and, in 1974 – after the Northern Cyprus invasion by Turkish troops – many more Greeks left (Keyder 1999). Nowadays, there are only a few dozen Greek-speaking people in Arnavutköy and 90 per cent of the population is of Turkish origin.

Arnavutköy is located on the European shores of the Bosphorus Strait between Kuruçeşme and Akıntıburnu and the larger area of Ortaköy, where the first Bosphorus bridge is located, and Rumeli Hisarı, where the European pylon of the Fatih Sultan Mehmet bridge lies. It belongs to the Municipality of Beşiktaş and it is a separate *muhtarlık* (local administrative district), a few miles from Taksim Square, which is considered to be the heart of Istanbul. It is a hilly area, built amphitheatrically from the top of the hill to the edge of a high-speed road that separates the buildings from the sea and the Arnavutköy İskele (ferry station).

The buildings are very close one to the other and its very narrow streets often lead to dead-ends (see Figure 1). Some of the streets are still stone-paved, a skill for which the Albanian residents of the area were famous. Arnavutköy still has a reputation for its Ottoman timber houses, some of which still exist today, contributing to the picturesque atmosphere of the area. However, in many cases the high cost of maintenance of such houses enforced the replacement of the timber parts with concrete – or the complete demolition of the house, and their replacement with a modern

8 The Treaty of Lausanne was a treaty signed on 24 July 1923 in Lausanne between the First World War Allies and the Grand National Assembly of Turkey, settling the Anatolian part of the partition of the Ottoman Empire. The treaty proclaimed the independence of the Republic of Turkey and the protection of the ethnic Greek minority in Turkey as well as the mainly ethnically Turkish Muslim minority in Greece. The Treaty also stipulated that much of the Greek population of Turkey was to be exchanged with the Turkish population of Greece. From the exchange were excluded the Greeks of Istanbul, Imbros and Tenedos as well as the Muslim population of Western Thrace.

construction. For example, in some cases most of the wooden part of the house has been replaced by concrete, maintaining only the shape of it. In other cases there are no wooden details left.

Arnavutköy is not a typical neighbourhood of what one might call Istanbul – a global city. There is a sense of closeness among the residents, expressed in their everyday activities. It is impossible to exit your doorstep without saying 'hello' to at least one person on your way to the grocer's (*bakal*), greengrocer's (*manav*), butcher's (*kasap*), shoe-repairer's (*ayakkabıcı*), or fisherman's (*balık satıcısı*). Many of the residents know each other, pay one another home visits, and meet at dinners or at the famous coffee-houses of the area. In short, there is a small-town feeling, which has been nurtured and strengthened especially ever since ASG was formed (1998), because of participants' efforts to raise consciousness for the Third Bridge issue.

The conflict

Until 1997 the debate of whether or not a third Bosphorus crossing (*3. Boğaz Geçisi*) was necessary was limited to the Ministry of Public Works and Settlements and certain bureaucrats such as the former Mayor of Istanbul (1984–9), Bedrettin Dalan,[9] on the one side and the Chamber of Architects on the other. The former claimed that the crossing was necessary whereas the latter claimed that not only it was not necessary but that it was also harmful as far as traffic congestion was concerned.

At the beginning of 1998 the Municipality of Istanbul (*İstanbul Büyük Şehir Belediyesi*) assigned the preparation of a traffic master plan for the city to Istanbul Technical University (*İstanbul Teknik Üniversitesi* – ITÜ). The master plan was delivered in December 1998 and among other solutions for traffic congestion suggested an underwater tube tunnel to improve public transportation. The construction of a Third Bridge was not part of

9 For the majority of scientists and Arnavutköy people I interviewed, Dalan is the synonym of 'disaster of the beauties of old Istanbul'.

it whatsoever (ITÜ ve İstanbul Büyük Şehir Belediyesi: 1998). However, in November 1998, a month before the master plan had been completed, newspapers published that the Ministry of Public Works and Settlements was planning the construction of a Third Bridge which would connect the European with the Asian sides of Bosphorus. The bridge would rest in Arnavutköy (European side) and Kandilli[10] (Asian side).

ASG was formed in 1998 immediately after the construction of the Third Bridge had been announced. Ever since, weekly meetings have been taking place, press releases have been printed, festivals have been held as well as dinners, tea parties and educational panels on the effects that the bridge would have on the neighbourhood's life. In addition, an oral history project of the area has been launched. All these activities aim at increasing the awareness of the area's important cultural and architectural history as well as of the destructive effects of the construction of the bridge. The arguments supporting this claim revolve mainly around the environmental effects of the bridge. Within the context of this discourse ASG involves concepts of global environmentalism such as sustainability, natural and cultural heritage preservation. It also incorporates issues of democratic participation and human rights.

Kousis and Eder (2001) claim that environmental politics is action taken less on environment than on society. It shapes not simply the environment but above all it shapes society. Thus any collective action which involves environmental issues in its agenda, it simultaneously involves societal issues. Therefore, I suggest examining the ASG initiative as a 'transenvironmental' protest. By the term transenvironmental I refer to a 'collective action that goes beyond the narrow environmental definition of the issue at stake allowing the environment to carry with it more meanings than just the narrow one of doing something good to nature' (Kousis and Eder 2001: 11).

10 Even though they have stated their support to ASG, the residents of Kandilli did not organise a similar protest. In an interview with the *muhtar* of Kandilli, she told me that the residents of Kandilli are close to Arnavutköy people but their small number does not permit a similar mobilisation.

Furthermore, following the definition given by Manuel Castells (1983) and Lowe (1986), I examine ASG as an urban protest, as its struggle is articulated with collective consumption demands, community culture and connected with the wider society through media, professionals and political parties. Throughout the course of its life, ASG rendered the issue public by having issued press releases and launching an Internet website not only within the country but also internationally. Foreign newspapers (see Chapter 3) published articles on the ASG struggle, ICOMOS (International Council on Monuments and Sites) listed Arnavutköy as 'heritage@risk', national and foreign NGOs were supportive to the campaign, scientists, politicians and celebrities declared their opposition to the construction of the bridge.

These developments are not the only aspects of urban character of ASG as a protest. The ASG struggle is a realm where various identities meet and reinforce the involvement with the campaign. In this sense, the urban face of ASG is evident through the participation of all the different subgroups within ASG that come together wishing to challenge existing public services (Lowe 1986). Therefore, in Chapter 3, I examine the different identities that co-exist within ASG. These identities form different subgroups with distinct characteristics which in turn compose ASG which consists of a core and a multilayered periphery. The core is constituted by the active participants, a group of nine people who are always present in meetings, events and are always informed about all the new developments on the Third Bridge issue. The second group of people, which I call less active participants, consists of five persons less involved but very strongly motivated against the construction of the bridge. Although their attendance to the meetings and the rest of the events is less often than the active participants, they are informed about any new developments of the group and the Third Bridge issue. The non-active participants are those persons who have never been systematically involved in the campaign but still support the cause. Finally, the fourth group is the group of the supporters. The supporters are not residents of the Arnavutköy but have participated in meetings and other events and have contributed to the campaign. Most of them are scientists and/or experts on traffic and architecture subjects and others are activists and celebrities. These four groups give ASG its particular character and render it a unique part of the Turkish transenvironmental movement.

In other words, this book is about a social movement organisation. I have chosen this definition, because I believe that a social movement does not only to oppose to certain decisions but also proposes different ways of doing things (Melucci 1989) like the Arnavutköy District Initiative does. However, I do not claim that ASG is a social movement but a part of social movements. More specifically, I define ASG as a social movement organisation (SMO) following the definition given by Resource Mobilisation Theorists; that is, a group that identifies its goals with the goals set by pre-existing social movements rendering itself part of them. In this sense, a social movement is not just a resisting group of people; it is rather a network of groups of people (Psimitis 2006) resisting on the basis of wider beliefs. Therefore, what constitutes a social movement is not the opposition to certain decisions but the alternatives it proposes. Thus, if one group resists the operation of a power plant and another one resists the construction of a dam or (as in our case) a bridge and proposes the same or similar alternatives, based on the same or similar beliefs, then all these groups together can potentially constitute a social movement.

In addition, I believe that such an examination can help the understanding of such groups in relation to those political contexts in which they are included. This claim stems from the belief that the very existence of such movements is a response to the existence of certain political decisions since the decision for the construction of a dam, a power plant or a bridge is a political decision per se. However, this book is not an ethnography of bureaucracy but an ethnography of the objections that bureaucracy entails.

Thus the present analysis revolves around three axes and forms an examination of ASG as a product of the Turkish public life. Consequently it discusses the differences and similarities that the initiative shares with other organisations and campaigns respectively. Finally, it focuses on the particular cultural character of ASG conspicuous, first through the formal presentation of the initiative, secondly through the informal accounts that embrace issues stemming from the evolving identities of its participants that mark the claims and the activities of the protest.

Outline of the following chapters

The book is divided into six chapters. The first (present) chapter is the introductory chapter, which offers a brief introduction of the conflict as well as a presentation of the methodology used either on the field or from other sites. Chapter 2 introduces Turkey and Istanbul as contexts in which ASG was born. Particularly in this chapter there is a brief historical account starting from the foundation of the Turkish Republic and extending until today and a brief presentation of Istanbul's architectural history. In Chapter 3, I elaborate on the ethnographic material which my research has provided me with and discuss the multiple faces of ASG. This chapter argues that even though at first glance ASG is a mere environmental group, in the ethnographic eye it reveals itself as a much more plural organisation both in terms of the issues it addresses and the people participating.

The issues addressed by ASG are issues met in other campaigns throughout Turkey, a fact suggesting that ASG is part of a network of protests which I call transenvironmental campaigns. Therefore, Chapter 4 explores the differences between formal environmental associations and, what I call transenvironmental campaigns. Chapter 5 constitutes a theoretical analysis of ASG consisting of a critical examination of the concept of civil society, an account on theories of social movements as well as a discussion on environmentalism. Finally, in Chapter 6 I sum up the main issues of the previous chapters and I reflect on more general topics for future consideration and research.

CHAPTER 2

Turkey and Istanbul: History, Politics and the Struggles of a Global City

Introduction

The study of cultural perspectives, according to Kay Milton (1996), is one of anthropology's distinctive specialities. In this pursuit, the context in which these perspectives emerge is important: 'Different locales, even different households respond differently within the same context, they are responding to similar forces and it behoves us to take the step beyond the household or the locales to understand those forces as a system so that we can better appreciate the variety of these responses' (Durrenberger 2003: 275). This chapter aims to introduce Turkey and Istanbul as the broader contexts which must be taken into account in research about the complexities of the Third Bridge issue. In order to do this, it is necessary to examine the historical and political processes which have formed this wider context.

Given that this book is about the conflict between the government and a neighbourhood initiative about an urban project in contemporary Istanbul, a brief presentation of the city's architectural history allows for the conceptualisation of the project as part of an 'urban milieu in which change is inherent' (Moore 1996: 24). In a larger, perhaps more significant scope, this chapter presents the urban history of Istanbul as a path to understanding contemporary societal change and structure (Hobsbawm in Çelik 1993: 18). This choice is based on the premise that the urban environment, according to Kousis and Eder (2001) 'is not only an ecological problem; it becomes a social (or even societal) problem at the end. It is a collective good that serves as a new medium in rearranging social relations between groups, thus rearranging relations of power and restructuring forms of social inequality.'

Following the establishment of the Turkish nation-state, Istanbul was neglected by a central administration which gave all its attention to the newly founded capital of the nation, Ankara. Through the years, Istanbul was increasingly subject to various types of alteration; some aimed at embellishing the city, others at serving particular needs of the citizens, others at serving both ends. What is of interest to the present work is the fact that the changes were 'beyond ordinary people's control' (Gulick 1984: 296) which, I propose, is the key to understand the opposition against the construction of the Third Bosphorus Bridge. As an extension of this point, I consider ASG as a protest-product of the effects that urban policies have on ordinary people. Moreover, since Arnavutköy is a neighbourhood of a global (Hamel et al. 2000; Sassen 2001) or world (Friedman and Wolff 1982; Hall 1996; Hannerz 1996) city – which seeks to establish itself as an economic metropolis with global ambitions – ASG can partly be seen as a result of the effects which globalisation has had on Istanbul at both the economic and the cultural level. Such contemporary metropolises constitute birth places of urban conflicts such as that exemplified by the struggles of ASG.

The Republic of Turkey (Türkiye Cumhuriyeti)

'*Ne Mutlu Türküm Diyene*' (Happy is he who says 'I am Turk')
— Mustafa Kemal Atatürk

The early years

Turkey was born from the remnants of the Ottoman Empire which, in terms of commerce, technology and industrialisation, lagged behind Western Europe. The Reformation and the Enlightenment, which had given rise to new forces of rational thought and scientific experimentation in the West, had limited influence in the East. According to some observers (e.g. Mango 2004) another reason for the weakening of the Ottoman Empire came from local nationalisms and the lack of a Turkish bourgeoisie with the capacity to support the creation of nation-state as did the bourgeoisie in the West. The Young Turks[1] Movement, an example of internal nationalism, resulted in the Young Turk Revolution (1908–9) which, in turn, led to the War of Independence in 1919 and the abolition of the Sultanate in 1922 (Mango 2004). The Republic of Turkey was established on 29 October 1923 as a result of the Independence War against the Allied Powers[2] which had begun on 19 May 1919 headed by a young Paşa (general) born in Thessaloniki (at that time part of the Empire, now part of modern Greece). His name was Mustafa Kemal, and on 24 November 1934, the National Assembly granted him the name Atatürk, that is, Father of the Turks (Mango 2002).

Following the declaration of the Republic of Turkey, a number of social reforms were enacted, mainly aiming at secularising public life and institutions. The reforms began in 1924 with the unification of education, completed in 1928 with the introduction of the Latin alphabet, which replaced the calligraphy of the previous writing system. In 1925, the fez (the religious head-cover) for men was replaced by the hat in both everyday life and as the conventional head-wear for state events, religious activities of sects were banned by law and the Western calendar was adopted by the state (Stirling 1993). The year 1926 saw the rise of the modern secular legislative system as opposed to religious law and the liberation of women as

1 The Young Turks was a movement of mainly military students that favoured constitutional reform during the monarchy of Sultan Abdul Hamid II. The Young Turk Revolution, which eventually led to a coup in 1913, constituted the foundation for Atatürk's revolution which subsequently led to the foundation of the Turkish Nation.

2 The main Allied Powers were France, the Russian Empire, the British Empire, and Italy. Their alliance was agreed during the First World War and sometimes they are also referred to as the Entente Powers. Later on many other countries joined the Allied Powers.

far as political and social rights were concerned. In 1928, the international numeric system was introduced and, in 1931, the metric system became used at the official and eventually common level. Finally, in 1934, religious attire in public was banned (Mango 2004).

The reforms were in accordance with what later would be called the 'Six Arrows' (*Altı Ok*), the six closely interlinked principles of Kemalism (*Kemalizm* or *Atatürkçülük*): Republicanism (*Cumhuriyetçilik*), Populism (*Halkçılık*), Secularism (*Laiklik*), Revolution (*Devrimcilik*), Statism (*Devletçilik*), and Nationalism (*Ulusçuluk*). Kemalism also stood for national sovereignty, national independence, national unity, peace at home and abroad, modernisation, scientific rationalism and humanitarianism. More specifically, Republicanism for Mustafa Kemal Atatürk was the only political form that would truly support the wishes of all people, men and women alike, including those who belonged to lower social classes, representing what Atatürk called Populism. According to Kemalism, Populism means that no individual, family, class or organisation is above any other; in fact, Atatürk often said, 'Turkey's true master is the peasant', signifying the importance of the ordinary people in the new Turkish Republic. Thus, Republicanism and Populism – as concepts and principles – are interwoven.

Secularism, perhaps the most important principle of Kemalism, meant not only the separation of state and religion but also freedom of thought, and in particular rational scientific ideas. The chief enemy of Kemalist secularism was considered to be Islam which, according to the Kemalists, opposed modernisation, democracy and progress. Revolution was designated as the principle most closely related to progress; it signified the modernisation of institutions according to Western bureaucracy, which was dependent on economic development – which, in the new Turkish nation, was regulated mainly by the state and the statist principle. Finally, the principle of Nationalism encompassed all of the others, in the sense that the Kemalist revolution was a nationalist action which secured the independence of the Turkish nation by consolidating territories and peoples who would need to be assimilated. In 1938 Atatürk died in Istanbul's Dolmabahçe Palace. The country would continue to at least the end of the Second World War to come to terms with the drastic changes he set in motion and with new Western values such as democracy.

Coping with democracy

In 1945 permission was granted for the foundation of political parties other than the Kemalist and by the 1950s a multiparty democracy replaced the single-party regime (Kandiyoti 2002). In the international arena, Turkey became a member of the UN in 1946, joined NATO in 1952, and (along with nineteen other countries) signed the Convention founding the Organisation for Economic Co-operation and Development on 14 December 1960. In 1964 Turkey was granted associate membership by the European Community, later to be named European Union (EU), but in the meantime, the multiparty democracy was interrupted by *coup d'état* in 1960. A second coup followed on 12 March 1971 aimed at halting the spread of left-wing culture and mentality in the country (Mango 2004). Mass imprisonment of young rebels and assassinations dominated the political scene until the 1973 elections. In 1974, a coalition government was formed by the secularist CHP (Republican People's Party) –the party founded by Atatürk – and the National Salvation Party (*Milli Selâmet Partisi* – MSP), a new party with Islamic orientations lead by Necmettin Erbakan. The same year, Turkey's invasion of Cyprus posed a problem to foreign relations, even though the climate was positive inside the country. The years that followed were marked by conflicts between right-wing nationalists and left-wing Marxists. Street battles were not uncommon between the conservative Grey Wolves[3] and leftist revolutionaries such as the one on May Day 1977, when right-wing extremists shot and killed thirty-four left-wing demonstrators (ibid.).

The economic situation of the country continued to deteriorate, and by 1979, political conflicts had developed into a virtual civil war. With the declaration of the third *coup d'état* on 12 September 1980, hundreds of thousands of people were imprisoned and hanged, and all the leaders of the political parties were incarcerated or banned from politics. Even as new political parties were formed, 20,000 out of 38,354 NGOs were closed down (Şımşek 2004) and the remaining unions, voluntary organisations and

3 Grey Wolves (in Turkish *Bozkurtlar*) is the youth organisation of the Turkish Nationalist Movement Party (MHP).

institutions were depoliticised (Beşpınar-Ekici and Gökalp 2006). Finally, another *coup d'état* came to interrupt once more the democratic process. This was because the winner of the 1995 elections was the Islamist Welfare Party of Necmettin Erbakan, who was the first Islamist Prime Minister of the country. His leadership was seen negatively by the National Security Council which in 1997 – with what the press called a 'post-modern or velvet *coup d'état*' – instituted new rules regarding religious freedoms. Political Islam was banned from public activities and individuals and companies associated with Islamist politics were penalised and punished. Politicians of the ruling party lost their office, including the present Prime Minister Reccep Tayyip Erdoğan (elected in 2002 and re-elected in 2007), who was imprisoned under the accusation of spreading religious fanaticism. In addition, the Welfare Party was banned (White 2002).

In spite of the efforts of the 'post-modern coup', Islamists had made their way well into Turkish politics, a fact supported by the victory of many pro-Islamist candidates in the municipal elections of 1994. Istanbul was among the municipalities which elected an Islamist mayor, none other than Reccep Tayyip Erdoğan, the leader of the Justice and Development Party (*AK Parti*) and twice elected Prime Minister. The emergence of Islamist politics in local elections in Istanbul did not prove to be what was expected from the, until then, dominant secularist discourse, as far as the cultural level is concerned (Navaro-Yashin 2002). Below, I will try to show that the politics regarding the development of Istanbul had followed a certain pattern either practiced by secularist or Islamist politicians. As an active participant of ASG put it in our interview regarding the Third Bridge:

> Ever since the issue was in the agenda, four governments have come to power and five ministers [of Public Works and Settlements] have been in the office. All of them support the bridge project. It does not matter whether it is a coalition, a secularist or an Islamist Minister. They all say yes to the bridge! Do you want to know why? [Because] There are interests behind it. Car companies, oil companies, you name it and you have it. They want to sell cars so they want to make roads to put them on (AP, a forty-year-old businessman).

Istanbul: A global city

The known history of Istanbul began about 300,000 years ago when the first inhabitants made their home in Yarımburgaz Cave. In 660 BC a group of Megarans (from Megara, a city near Athens), after consulting the Oracle of Delphi, founded Byzantium, which became part of the Roman Empire in AD 73. Around AD 324, Byzantium began to blossom when Constantine the Great decided to move the capital of the Roman Empire to the East, that is, to Byzantium. He first called the new capitol Nea Roma (New Rome) and on 11 May AD 330 he named it Constantinople after himself; it became the centre of the Byzantine Empire. After almost fifteen centuries of Byzantine and Ottoman architectural development, during the nineteenth century foreign capital came into the region, and the modernisation as well as the reformation of the Ottoman Empire began (Çelik 1993). A good number of mosques were built in baroque style, such as the Ortaköy and Dolmabahçe Mosque on the Bosphorus banks, and in 1853, when Dolmabahçe Palace was completed, Topkapı, the shining star in architectural achievement, made way for the grandeur of the new palace. The Çırağan Palace was built in 1871 and during the same period exhibition centres, theatres and buildings were constructed, among them the Yıldız Kiosk, a Western-style country mansion, and the Şale Kasır, inspired by the Swiss chalet.

Today, Istanbul is the largest city of Turkey with 12,573,836 inhabitants (http://tuikrapor.tuik.gov.tr/reports/rwservlet?adnks=&report=turkiye_il_koy_sehir.RDF&p_kod=1&desformat=html&ENVID=adnksEnv). It is situated in the north-western part of Turkey and lies between the Marmara Sea to the south and the Black Sea on the north. The original city was surrounded by seven hills with steep slopes and ample summits. Istanbul is the only city in the world located on two continents, Europe and Asia. The Bosphorus Strait separates the European, west side of the city from the Asian East, and is the only seaway from Black Sea to the Aegean Sea and the Mediterranean. The European side of Istanbul is also divided by a waterline, the Golden Horn, which divides the Old Istanbul (south) and the New Istanbul (north). Some 36 per cent of the total population of the city lives in the new Asian side of the city lives, whereas 64 per cent (Bliss n.d.) lives on the European side The old city is well known for its very old buildings in narrow streets and many important historical buildings such as Topkapı Palace, Haghia Sophia, the hippodrome, the Grand Bazaar and Istanbul University. However, the old city is also characterised by the squatter settlements or *gecekondu* (literally meaning 'built overnight') which surround it and consist 65 per cent of all buildings in the city (Yalcintan and Erbas 2003).

In the heart of the new city we find the Beyoğlu district with Taksim Square and the famous shopping street Istiklal Caddesi (Independence Street), high-rise buildings (apartment blocks, business buildings and hotels, consulates and banks) and the Galata Tower, which is considered one of the landmarks of Istanbul by both visitors and inhabitants.

Üsküdar and Kadiköy are the two main suburbs of the Asian side. Baghdad Street (Bağdat Caddesi), one of the most famous shopping streets of Istanbul, is full of apartment buildings, where most of the residents of this side of the city live. In addition, the Asian side is inhabited by upper-middle-class families who live either in newly built houses or in old timber villas, which are called *yallıs*, located by the Bosphorus shores. By the Bosphorus Straits we also find a few mosques, such as the Ortaköy Mosque, the Fortress of Europe and Asia, the Dolmabahçe Palace and fishing villages which extend to the north until the Black Sea.

The Princess Islands (Adalar) are located about 20km south-east in the Sea of Marmara; they belong to the Municipality of Istanbul. They are a complex of four islands (Buyukada, Hebeliada, Burgazada and Kinaliada) that used to be the burial place of Byzantine emperors and princesses. Later on, these islands were developed as summer resorts for wealthy Ottomans and today they are one of Istanbul's most popular tourist destinations.

Today the greater Istanbul area is home to manufacturing plants, which comprise 35 per cent of the country's manufacturing industry. In addition, automobiles, concrete, cigarettes, fruits, olive oil, silk, glass, cotton, leather and pottery are produced in the peripheral areas of the city. Istanbul is the largest port in Turkey, thus shipping is a major source of income; the city is a main financial centre as well as a top tourist attraction (Bliss n.d.).

Istanbul in the early Republican years

After the declaration of the Turkish state, one of the first things Kemal Atatürk did was to establish Ankara as the capital of the nation. Istanbul had already begun to lose its glory before the title of the capital passed to Ankara. According to Mango, by 1923 Istanbul had become provincial and Atatürk's administration did not alter this image. Only a few changes took place such as the rise of new blocks of apartment buildings, monuments of Kemal Atatürk and the building of his summer house on the Florya Beach, close to the airport. By 1938, when Atatürk died, Istanbul had become an old city and the city's population was getting around by trams built by foreign enterprises (ibid.). Only a few private cars were in circulation in the narrow streets of the city as well as using the Unkapanı (or Atatürk) Bridge, a pontoon bridge which was completed by a French firm in 1939.

The Marshall Plan and the 1950s

Even though Turkey had managed to remain neutral during the Second World War, it was included in the list of nations to receive financial aid under the Economic Assistant Act or as it was better known, The Marshall Plan. The plan was presented by the US Secretary of State George C. Marshall in 1947 as solution to the catastrophic consequences from which Europeans suffered due to the Second World War. It suggested that the US provide financial aid to stop hunger, poverty and desperation in Europe and revive a working economy in the world so as to permit the emergence of political and social conditions in which free institutions could exist. In other words, the plan aimed at stabilising the international order in a way favourable to the development of political democracy and free-market economies. Subsequently, the prevention of the spread of communism in Western Europe was also in the agenda. A result of that proposal was the Economic Assistant Act (EAA) signed by President Harry S. Truman, who enacted the plan (3 April 1948). Almost all European countries, except for those of the Soviet bloc, were part of the plan, including Turkey. The

American Congress appropriated US $13.3 billion for capital and materials to help rebuild Europe's economy. Furthermore, the plan provided goods, created trading partnerships and extended the administration of the American policy into areas outside the United States (http://loc.gov.exhibits/marshall).

As Keyder mentions (1999: 12), the postwar period of national development in Turkey was heavily regulated by political decision making and relied on strict control over imports, foreign investment and international exchange. The Marshall Plan aid was supervised by the donors (i.e. the Americans) and as an article entitled *How to Do Business under the Marshall Plan* in *Kiplinger* magazine (a publication for financiers) stated: 'The Marshall Plan is very much a business plan.' Hence the receiving country (Turkey in this case) had to be accountable to its donor, and in response, the Americans 'created a plan for the construction of Turkish roads and contributed to the creation of the Turkish Highways Department' (Mango 2004: 44). The then prime minister, Adnan Menders, favoured the construction of large boulevards – such as can be seen today in Istanbul – where private cars but not public transportation vehicles could circulate easily (Keyder 1999). His vision was that 'Turkey would become little America' (Yalçın 2002).

During the 1950s Istanbul began to experience a rapid population growth due to internal migration and its architectural landscape started changing drastically. *Gecekondus* began to mushroom and, by 1960, the city's population had risen to 1,500,000 inhabitants, double the 1938 figure (740,000). Private car ownership increased, too; many modern buildings began to be constructed and unplanned architectural growth continued until the 1970s. The number of cars increased and the need for new crossings over the Bosphorus began to emerge. For example, the Golden Horn Bridge was erected in 1974 and financed by Japanese Credit as part of a long-distance expressway network connecting Asia and Europe (Masashiro, Toshimitsu and Mitsubiro n.d.). This network included the construction of the two bridges across the Bosphorus in the early 1970s and late 1980s. Even though the decision to build a bridge across the Bosphorus was made in 1957, when Adnan Menderes was the prime minister, the contract was signed with the British firm Freeman Fox and Partners for TL 303 million in 1968, and the construction of the first bridge started on 10 February 1970. It was completed in 1973 and the bridge was named after the Strait; i.e. the Bosphorus Bridge (*Boğaziçi Köprüsü*) (http://adayinlife.typepad.com). It is a suspension bridge mainly used by private cars as well as public transportation buses, with the tariff for private cars being YTL 4.00.

The plan for a second bridge was assembled as early as in 1977, four years after the first bridge was constructed. The initial plan, prepared by the British construction company Freeman Fox and Partners, was designed to accommodate five bridges. The first one would connect the Rumeli Hisarı and Anadolu Hisarı areas. Initially it was planned as a double bridge in the shape of a delta. Its second part would be the second bridge. The third and fourth bridges would be constructed between Arnavutköy and Vaniköy, areas which were also designed to accommodate two sections. Finally, the fifth bridge was designed to be constructed between Emirgan and Kanlıca areas. So far, one of those bridges has been constructed, the Fatih Mehmet Sultan Köprüsü (Fatih Sultan Mehmet Bridge), named after Fatih the Conqueror, which was completed in 1988. Like the Bosphorus Bridge it is a suspension highway bridge and the tariff costs the same.

In summary, beginning in 1948 Istanbul developed according to a foreign Western technocratic mentality imported through capital and expertise. In terms of transportation, choices for development favoured large highways (rather than railways), designed to accommodate private vehicles, which connected to the Bosphorus crossings. Currently, mass transportation on these highways and bridges is limited to public buses without pedestrian or bicycle/motorcycle lanes.

Istanbul in the era of globalisation

The new Millennium and the Marmaray project

Mass transportation was improved during the 1990s when the Istanbul Metro was constructed; initiated in 1992, the first line was completed in 2000. The 1990s were the decade of the Islamist-oriented mayors of Istanbul (Reccep Tayyip Erdoğan and Ali Müfit Gürtuna), who were more concerned with improving the city's social life and cultural politics (see Navaro-Yashin 2002) than on focusing upon infrastructure needs. The public policy agenda of Reccep Tayyip Erdoğan gave high priority to environmental issues such as the improvement of water distribution to Istanbul, the publication of books on environmental issues and the encouragement of well-known environmental activists to present their views (Özdemir 2003). The decade was marked by the 1994 and 1999 economic crises, which continued into the new millennium, a fact that did not permit large infrastructural developments. Nevertheless, in 2004 one of the largest and most ambitious constructions in the history of the city was inaugurated: the Marmaray Underwater Tunnel. The idea of a railway tunnel under the Bosphorus Strait was officially considered for the first time in 1860, along with a number of technical impediments which came to attention in both architectural and lay suggestions. Some investigations advised that it would be impossible to allow a tunnel to be placed on or under the seabed; other designs advocated a 'floating' type of tunnel which would be suspended on pillars sunk deep into the seabed.

The necessity of a railway mass transit connection from west to east in Istanbul and under the Bosphorus Strait gained momentum in the early 1980s, and the first comprehensive feasibility study conducted in 1987 concluded that such a connection would be feasible and cost-effective. The project was discussed during the following years, and in about 1995 it was decided that additional feasibility studies should be more detailed and updated. These studies were completed in 1998, and the findings agreed with the earlier conclusions that the railway transit connections would offer many advantages to the people working and living in Istanbul as well as ease the increasing problems of traffic congestion in the city.

A loan agreement signed in Ankara on 19 August 1999 released US $117 million from a total funding of US $866 million provided by the Japan's Overseas Economic Co-operation Fund. Unfortunately, the 2001 economic crisis prevented state funding and the project was slowed down. In 2003 and 2004 discussions were held with European Investment Bank (EIB) to provide major portions of the funding for those projects. Principle agreements to fund major portions of the Commuter Rail Systems were made in autumn 2004 (http://www.marmaray.com). Finally, on 9 May 2004 Prime Minister Reccep Tayyip Erdoğan inaugurated the Marmaray Underwater Tunnel Project. The entire undertaking included plans for a 13.3km Bosphorus crossing and the upgrade of 63km of suburb line to create a 76.3km high-capacity line between Gebze and Halkalı. Access to the tube would be by tunnels bored from Yenikapı on the European side and Sögütlüçesme on the Anatolian side, with intermediate stations at Sirkeci and Üsküdar and an interchange station with the Istanbul Metro at Yenikapı. The line capacity will be 75,000 passengers per hour in each direction.

The inauguration of the Marmaray Tunnel was a highly publicised event. Many members of the cabinet, the Prime Minister as well as mayors of foreign capitals and ministers of transportation from other countries attended the ceremony. Supporters of the government gathered to demonstrate their content for the initiation of the project. ASG participants were also present in an effort to declare their support to the Marmaray Project as clearly preferable to the Bridge Project. In this way they situated themselves in favour of an urban development goal which gave priority to improved public services such as mass rather than private transportation, an issue in many contemporary large metropolises.

Local protests in global cities

The 'city' as a collectivity of people and a geographic location has been studied by social scientists from different and often heterogeneous backgrounds. Political scientists, economists, geographers, sociologists and social anthropologists have shed light on life in cities as a distinct field of study from a variety of theoretical perspectives. Urban anthropologists often focus on the complexity of social life in cities (Sanjek 2004) looking at the daily life of migrants (Hannerz 1983), urban poor, work environments, voluntary associations, and social movements (Nash 2005). Drawing on theories of dependency, the world system, globalisation and Marxist perspectives, urban anthropologists shaped a distinct field of study which 'officially' emerged in 1972, with the publication of the first issue of the journal *Urban Anthropology*.

Those following the dependency theory model study the ways in which large metropolises of the developed nations dominate the economies of cities emerging in developing countries. For example, Kutsche (1989: 130) perceives cities as 'centers of power that control semi-peripheral centers, which in turn control other centers along a chain of indefinite length, ultimately controlling satellites'. The Wallerstein model of the World System gained supporters from anthropology who investigate the influence of developed cities not only on other cities but in general across the globe. Hannerz (1996) for example, discusses the transnational connections created by world capitalism where, according to his perspective, the globalised world is a product of the dialectic interplay between local and global forces. Having conducted fieldwork in urban settings, Hannerz claims that cities are places where the intensity of this interplay is larger and thus more easily observable. The Marxist point of view favours an analysis of the city as a process (Moore 1996). Urban sociologists such as Walton (in Ho Kwok-Leung 2000) suggest that a city combines market, political authority and community. City life is a process of the interplay of economic forces, political control and community interactions (Ho Kwok-Leung 2000: 4); thus, any theory of the city must begin with an examination of social conflicts between these forces (Castells 1983: 318).

The work of Manual Castells, one of the most influential theorists in contemporary urban studies, centres on urban change which, he claims, is powered by the interests and worldviews of social movements. According to Castells, urban forms and functions are produced and managed by the interaction between space and society, that is, by the historical relationship between human consciousness, matter, energy and information (Castells 1983: xi). He goes on to assert that the city is defined by social interests and values, where the dominant ones have been institutionalised and, thus, resist pressures to change coming from primarily grass-roots mobilisation. In this view, urban transformation is a result of the interaction between dominant interests and grass-roots resistance. Finally, for Castells, social change is the product of class struggle, the autonomous role of the state, gender relations, ethnic and national movements as well as movements that are self-defined as citizen movements.

Similarly, Sassen (2001) claims that global cities host the conflicts and the battles which previously took place in colonies. These new battles occur between the new transnational professional class (the new city users who treat the city as a transterritorial environment) and the underpaid immigrant working class (which provides the material conditions for the corporate world of power). The large numbers of the latter in global cities gives them a strategic role for laying claims on the city, and allows them to negotiate the conditions of their powerlessness as well as to develop a form of politics which challenges the project of the global elites. In this sense, urban social movements can be seen as products of globalisation and more specifically, as responses to economic globalisation that overlooks or demotes locality. This is what Hamel and colleagues (2000: 6) suggest when they claim that 'globalisation is intrinsically linked with oppositional cultural activism'. The costs of globalisation are enormous for the biggest part of the population of global cities (Sassen 2001); as one result, urban protests continuously emerge.

Many examples around the world confirm Sassen's claim (see Introduction). The Third Bridge case as well as the Beyoğlu demolitions (see Chapter 4) are two of the most characteristic examples of such protests in Istanbul. As Lewellen (2003) claims, any anthropological definition of globalisation would also have to include the local-level resistances and adaptations to these processes in which the urban political economy becomes less important than the micro-social processes that give birth to urban movements. The social space of operation of such movements is the 'extra local' and this defines the difference between urban and other movements. This difference is located to their perspectives of institutions of local development and management when confronted with patterns of domination reproduced by political elites (Hamel et al. 2000). Similarly, Castells (1983: xvi) points out that:

> In order to understand cities and citizens it is essential to analyse the relationship between people and urbanisation. A way to study this relationship is by studying mobilisations that aim to change the city. Thus, the examination of urban movements suggests that they are collective actions consciously aimed at the transformation of the social interests and values embedded in the forms and functions of a historically given city.

Urban social movements, as groups which claim rights to public services and explore new meanings for cities, challenge dominant cultural values and political institutions and refuse the existing spatial forms. In this sense, 'urban movements present an opportunity for mediating crises within the system. They represent a bifurcation in the system and they mean to fill this space by pointing to the system's most glaring crisis, the flow of capital and its effects upon the communities in which people live their lives' (Wallerstein in Castells: 5).

Big cities in Turkey – such as Istanbul, Ankara and Izmir – attract human resources through their concentrations of capital which resulted from the liberalisation of the economy. Today 72 per cent of the Turkish population lives in big cities (Ignatow 2007). In Istanbul, with its global character, the high priority on efforts to attract foreign investment and development projects relevant to this goal (ibid.) resulted in the construction of business districts and the support of to facilitate access to them. As Ayşe Öncü (1997: 57) writes about Istanbul: 'The internationalised business centre towards the north of the Golden Horn, with its deluxe hotels, modern office towers and wide avenues, was to host global functions, welcoming conventions, businessmen and tourists.' As a result, internal migration to Istanbul is massive. Every year 500,000 migrants come from rural areas to reside in the city (Yalcintan and Erbas 2003). According to the 2000 census, the urban population growth rate is 32.6 per cent and the proportion of urban populations to the rest of Turkey is 59.25 per cent (TURSTAT 2004).

This massive internal migration resulted in the creation of a global city[4] organised by social relationships in both local and transnational contexts. As Hannerz notes, global cities are constituted by four categories of people: transnational entrepreneurs in banking and finance who provide legal services, accounting, technical consulting, telecommunications and international transportation; the research and higher education people; the third-world populations; and the people concerned with the promotion of 'culture and tourism'. Dominant classes in these cities are the transnational elites and thus world cities are structured in a way that serves their lifestyle (Hannerz 1996). This is exactly the assumption of ASG, a forty-four-year-old high-school teacher and active participant of the group, who stressed: 'The bridges serve only one out of fifteen million individuals who reside in Istanbul. Those individuals live on the Asian side, work downtown and move with their private cars.' As a result many people are displaced, and the areas not used to house business-related buildings host unattractive functions of the developments or are abandoned (Mayer 2000).

The process continues when metropolises such as Istanbul undergo privatisation and individualisation of services and activities which were previously state-funded (Nash 2005). In the name of free enterprise, neo-liberal political actors assign the employment of public resources to the private sector in order to make their cities more competitive in the global arena (Kentor et al. n.d.). Thus, the state seems to withdraw support from public services which are, in turn, privatised, a trend which marks the current situation of the global political economy. The accounts of my informants indicate that the same kind of neo-liberal politics have been

4 By the term 'global city' I refer to cities highly integrated in the worldwide circuits of capital, acting as control centres of the global economy (Friedman and Wolff 1982) and shaped by transnationalism and new technologies (Hamel et al. 2000: 7).

followed in Istanbul. A university professor involved in the ASG campaign mentioned:

> Public transportation in Istanbul was better before. People used to move with the ferries that were cheaper, cleaner, much more comfortable. If there is good public transportation then people will use it and I still don't understand why they don't focus on that. I mean, building a bridge is much more expensive, if you use the sea is less expensive. You build the *iskele* [dock] and all you have to do is buy a few boats and you have them running on a more regular basis, every half an hour let's say. It's going to be much cheaper and in the long run it is going to be much better.

As global metropolises, world cities are affected by the currents of economic globalisation to a larger and more systematic degree than cities less integrated in the global processes (Mayer 2000). Istanbul is part of such processes in various ways. To continue with the comments of the above informant:

> The tankers crossing the Bosphorus dump their garbage in the sea and what happens is because of the currents that come very quickly, a lot of the rubbish comes to the surface. And because they built this pyloned road above the sea, the rubbish gets underneath. This really makes a terrible ... offensive collection of rubbish on the side. Beşiktaş municipality has a special boat that gathers the rubbish but of course it is very difficult because the rubbish goes underneath the pylons. There was a tanker accident about three or four months ago. Süleyman has some very good slides of what happened. Ten days for the people to clean the village. There was petroleum and it came to the cost of Arnavutköy. And you know people are fishing there ... It's one of the most favourite places. It really created a great pollution. And the frequency is big.
> [Question: Why is this allowed?]
> What happens is that sometimes the boats are stopped and controlled and are given a penalty. But there are international laws. You cannot give too much penalty.

Thus, the global-scape in which urban conflicts and movements emerge may constitute the basis of analysis between the functions and activities of the global cities and the issues and actors of urban social movements (Sassen 2001). The emerging conflicts in such cities have global character in the sense that the economic, political and community forces are shaped by transnationalism. The ASG case is an example of such conflicts which will be analysed in greater detail in Chapter 4.

'No to the interests underlying the bridge':[5] The struggles of a global city

Istanbul has changed a lot through the centuries. As the capital of the Byzantine and Ottoman Empires, it was a location for monumental architectural projects – great palaces, churches, mosques, walls, bridges, and so on. During the nineteenth century the wish of the Ottoman rulers to assimilate Western elements in architecture and administration changed the city again. The creation of the Turkish nation gave new meaning to urban landscape. Big boulevards, new bridges and highways were constructed. As Turkey has integrated itself into Western capitalism and more recently (from the 1980s onwards) into economic liberalism, it has also established a relationship of interdependence; first, through import of aid and expertise and, secondly, through international loans and exports (Karafotakis 2000). Some bridge projects were externally funded as well as the underwater tunnel project. As the ASG claims, the Third Bridge project is a product of underlying interests; interests that have to do with maximisation of economic profits. To a certain extent the governments' opinion concurs with this claim. In response to my question, 'Why does the Ministry of Public Works and Settlement insist on having the bridge built in Arnavutköy?' a high-ranking Ministry bureaucrat replied:

> First of all we have to protect our environment. Building a bridge in the northern part of the Bosphorus where all the water reserves of Istanbul are situated would be criminal. Therefore this alternative has been rejected. Our second choice is Arnavutköy because the distance between the two parts of the city [Asian and European] is small and the construction will cost less than in other parts where the straight is not so narrow.

Of course this was not ASG's allegation, but as an active informant implicitly put it, 'There is a lobby behind the bridge. Car companies mostly. We call it the "black lobby". The tunnel lobby we call the "white lobby".'

5 ASG placard.

The comments indicate that ASG members acknowledge that there are economic interests behind any project, even behind the project they support. One of the main points the press representative of ASG stressed in one of our discussions was the following:

> If you want to study our protest, you have to look back. We are not just reacting to the bridge. We are reacting to a series of policies implemented in this country, in our city [Istanbul]. These policies started more than fifty years ago, when Turkey decided to receive US financial aid; you know, the Marshall Plan.

For my informants the Bosphorus bridges and the Third Bridge are not simply undesirable development projects; they are symbols of Turkish obedience to foreign donors and big capital. As explained by a teacher of foreign languages who resides in Arnavutköy and participates in ASG: 'At some point those who rule this country must understand that people, all Turkish people, should be heard and their opinion should be a factor to their decision making.'

Unfortunately for ASG, Istanbul appears to be highly integrated in the world political economy and in processes transcending the national context which relate to economic forces, uninformed of ordinary citizens' concerns and needs. As mentioned above, from the nineteenth century Istanbul began to receive foreign investments that were regulated by political decisions. In the second half of the twentieth century the incoming flow of foreign capital took the form of Marshall Plan aid, which was used under the supervision of the donors. In fact, the Department of State Highways – one of the main actors inf the Third Bridge conflict as well as in the construction of the other two bridges – was founded under the guidance of US experts. One of its purposes was to make sure that the distribution of incoming funds for building Istanbul's network of boulevards and peripheral highways would be distributed in an official and accountable manner.

Another fact suggesting the transnational character of processes taking place in Istanbul is that the Bosphorus, which is one of Istanbul's most essential assets, is the only maritime overland link from the Black Sea to the Mediterranean; it allows Istanbul to be an important actor in international trade between the two regions. This international role of the Strait is prioritised over the domestic use of it. The number of ferries which cross the Strait has decreased and currently many residents now use their private cars or public buses to cross the bridges, creating considerable traffic congestion during rush hours. The university professor active in ASG mentioned earlier claimed that:

> All the governments have tried to change the city according to the needs that car usage creates, whereas the opposite should have been done ... There is a car-industries lobby behind political decisions which do not concern only Istanbul or Turkey. The attempt to increase speed limit, for example, relates to the effort to render cars more competitive with the rest means of transportation. Car purchase in Europe decreases and this is preoccupying for the car industries.

As Paul Durrenberger (2003: 276) maintains:

> The states in serving the interests of corporations are unable to serve the interests of their citizens by protecting their environments or insuring their economic welfare. In democratic states, those in which citizens elect governments, this causes tensions. There is a tension between the interests of corporations and interests of populations. That is what we see playing out in the process of globalisation as numbers of people gather from around the world to protest wherever international bodies meet to discuss policies of world trade. If we want to understand these movements and their manifestations from protest to suicidal attacks, we must understand the system that gives rise to them.

Conclusions

In his book entitled *Istanbul* the Turkish Nobel Prize-winner Orhan Pamuk (2005) compares the feeling of *tristesse* described in Levi-Strauss' book *Tristes Tropiques* with the feelings of a visitor to Istanbul. According to Pamuk, this feeling is different from that of sadness, inherent to the residents of Istanbul. Anthropologically speaking, the sentiment of *tristesse* falls under the category of *etic* when referring to feelings of the Western traveller for this city, possibly associated with Western guilt towards a city within a developing nation. The sentiment of sadness falls under the category of *emic* feelings of the natives, who are deeply unhappy about their city's decadence, perhaps even grieving about the loss of its past grandeur. Pamuk claims that the residents of Istanbul are sad because they live among the destroyed relics of a city which had been the centre of great and glorious empires. To confront this sadness, Istanbulites deliberately ignore those monuments and, in many cases, they destroy them or wish them destroyed so that they are not reminded of their city's past and lost glory.

Pamuk's opinion on the deliberate destruction of the city's Roman, Byzantine and Ottoman past due to sadness for past glory has not been one of the topics of this book, nor can it be substantiated from the data produced during my fieldwork. However, such an approach is interesting since the destruction of old Istanbul underlies a complaint I heard many times while I was living in the city, both from my informants and other people, like taxi drivers, architects and others. As one of my informants put it, 'Istanbul is as unique, as is the way it is being destroyed.' I cannot say whether the obliteration of old constructions is the result of state policy or the acts of private citizens in an attempt to avoid reminders of the glory they no longer possess. This could be a subject for further anthropological investigation. I can say, however, that after the establishment of the Turkish Republic Istanbul, which lost its title of the capital to Ankara, has been neglected by official policy and has become subject of unplanned modernisation and growth.

As mentioned earlier in this chapter, contemporary Istanbul is a global city which combines financial districts with skyscrapers, old mosques and churches, museum-palaces, apartment blocks, *gecekondus*, small streets and big boulevards, bridges and peripheral highways (*çevreyolu*) crossed by millions of cars, a small tram-line and a small metro-line. It seems that ever since 1923, the city's administrators have been determined to modernise it at any cost by destroying old neighbourhoods or building over them (see Chapter 4 for the Beyoğlu case), and making Istanbul a financial centre. In addition, it seems that this technocratic mentality has survived until today. As a result of this mentality, many of the modern constructions are now considered to be the city's landmarks. The Bosphorus bridges are two of the most characteristic examples. Many postcards of Istanbul picture the two bridges in a picturesque way, implying that they are one of the city's charms. The international tourist campaign for Istanbul for the summer period of 2007 sponsored by the Turkish Ministry of Tourism is almost completely based on the Bosphorus bridges, presenting them either as the city's modern achievements or as the city's organically evolved development. As one high-ranking official of the Turkish Highways Department told me: 'The bridges are necessary and that is why they were built. Istanbul needs more roads, so that people can have easy access to their work. If fifteen houses in Arnavutköy must be demolished for the greater benefit of millions of people who need to go to their work quicker, so may it be!' A similar statement, but in a more intense tone, was made by the head of the State Highways Department in 2001 during an interview to Milliyet Newspaper: 'We will construct the Third Bridge no matter what!' (*3. Köpru yapacağız!*).

Yet, the policies that brought about those changes in the past and the globally informed policies of today have not only great achievements to be proud of. The transformation of Istanbul to a financial centre has attracted internal and foreign immigrants who need somewhere to live and the fact that Turkish welfare state does not provide housing for these categories of citizens has resulted in poorly constructed, unattractive big apartment buildings and *gecekondus*. The construction of large boulevards and the Bosphorus bridges, without simultaneous improvement of public transportation has resulted in traffic congestion, air pollution and displacement.

Following this train of thought, the book examines ASG as part of what Falk (1993: 39) named 'globalisation-from-below'; that is, transnational social forces animated by environmental concerns, human rights, hostility to patriarchy, and a vision of human community based on the unity of diverse cultures seeking an end to poverty, oppression, humiliation, and collective violence. All these elements lead to the need to examine the ASG a mobilisation, such as all mobilisations which are inherent parts of global cities and as a result of political choices which affected Turkey in general and, Istanbul in particular. Indeed, as I will try to show in the next chapter, the anti-bridge campaign articulates environmental and urban with political demands.

CHAPTER 3

Arnavutköy District Initiative: Environmentalism, Transenvironmentalism and the Identities of an Urban Protest

Introduction

The purpose of this chapter is primarily to analyse the environmental discourse that ASG uses, and the elements that give ASG its transenvironmental character. The environmental aspect is evident from the formal discourse used to oppose the construction of the bridge, a discourse which 2ough social activities such as the ASG annual June festival, tea parties and grand dances. Such claims and activities are associated with the identities of the participants of ASG. I refer to identity in the plural because within ASG there are multiple identities stemming from different demographic characteristics and beliefs which define different subgroups; for example, identities may be secular, leftist, local and transnational.

As will be discussed below, ASG is a group of people consisting of a core and a mulitlayered periphery. The core consists of the active participants (AP), whereas the periphery consists of the less active (LAP) participants, the non-active participants (NAP) and finally the 1ch, along with its activities, mark the identity of the group which is multifaceted and multilevelled.

ASG as an environmental group

One of the central arguments ASG raises against the construction of the bridge concerns the potential environmental damage in the area. As stated in the initiative's website (www.arnavutkoy.org), 'The historic, cultural and natural assets of both sides of the Bosphorus are going to suffer either directly or indirectly because of the proposed bridge.' One of the ASG press releases begins: 'Dear Minister, Bosphorus and Arnavutköy is protected by the Cultural and Natural Assets Protection Act and this is why we want the decision of the construction of the Third Bridge cancelled.' The ASG website lists these assets as:

a. Thirty-eight monumental constructions
b. 292 examples of civil architecture (houses)
c. Five natural green conservation areas
d. Forty-two *yalıs* or waterfront houses from the Ottoman era
e. Thirty monumental trees
f. Retained walls, garden walls (Decree 9483 of The High Council of Monuments).

The participants of ASG also claim that, according to scientific studies, air pollution caused by cars crossing the existing bridges of Bosphorus is enormous.

In addition, as will be seen, the ASG makes use of sustainability and global cultural heritage arguments to influence its audience. For example, a press release on the website states: 'Arnavutköy is one of Istanbul's oldest neighbourhoods, many of which are being lost and destroyed. This is why the protection of Arnavutköy is of great importance.' This message is reflected in one of the often-used arguments of my informants, namely that Arnavutköy should be preserved for future generations. In agreement with this, five other participants from all the identity subgroups stated:

> Arnavutköy must be preserved for our children. If you demolish the houses or the physical environment, they will not have the chance to create a connection (*bağ*) with the past (NAP, forty-five-year-old architect).

> I do not so much care about my house. I care about the sea. The sea should not be destroyed. My children should be raised by this very sea (AP, forty-year-old businessman).

> If someone could prove to me that the bridge can solve the traffic problem of Istanbul then I would definitely support it, but all is going to do in the long run is to destroy this beautiful place, which should remain as it is (LAP, forty-two-year-old psychologist).

> If Arnavutköy is destroyed, which it will if the bridge is built, hundreds of years of history will be destroyed (LAP, fifty-year-old architect).

> I am planning to leave Arnavutköy but I fight against this bridge because this will always be my neighbourhood. This area should stay as it is for future generations (AP, sixty-year-old pensioner).

Statements such as the above suggest that the arguments for the preservation of Arnavutköy follow a rhetoric of sustainability. Protect it in the present, in order to use it in the future; and it is not only to be used by the locals but for everyone because:

> ... Arnavutköy is very old. Its history is very rich and important not only for Istanbul but for the whole world (AP, forty-four-year-old high school teacher).

> Arnavutköy does not only belong to us. It belongs to the global cultural heritage (AP, forty-year-old tourist guide).

> Nowadays worldwide, the cultural heritage is more important than any other public interest. Nobody can destroy a district, building or a monument. We have made a lot of steps and that is why it is very difficult for them to build this bridge (LAP, fifty-year-old architect).

Clearly Arnavutköy fulfils the criteria of a global site; consequently, its protection lies in everyone's hands. The website of the Greek Fellowship of Mega Revma founded by Christian Arnavutköylites who were expelled or voluntarily left the area during the 1950s and the 1960s crises explains the global scope (as well as the cultural uniqueness) of the neighbourhood:

> Arnavutköy is one of only a handful of neighbourhoods where the cultural mosaic of Greek, Armenian and Turkish influence, once so much a part of Istanbul life, can still be seen and felt, both in the architecture and in the residents. It is one of the last areas where residents and tourists (many of which visit Arnavutköy every year), can see at first hand the famous wooden-house architecture which characterised the rich style of the late nineteenth and early twentieth centuries. It is one of the last areas to support a functional Greek school and church, and contains the ruins of an ancient, and yet undocumented synagogue and other sites of architectural and cultural importance. Arnavutköy is indeed a site of cultural heritage (www.megarevma.net).

For ASG participants and supporters, Arnavutköy represents an existing ideal place and we have to act toward its protection because this is proper for our society and besides, 'the need to believe in paradise is universal' (Argyrou 2005: 85). As the online declaration of the ASG continues:

> We believe that the fate of Arnavutköy, and the disregard with which the government approaches this important historical site, ultimately reflects to Turkish citizens, visitors, and architectural historians everywhere, the fate of our society as a whole. For this reason we call on you to help us defend this beautiful and important place from the wanton destruction which this recently unveiled plan threatens. Specifically, we call upon you to voice its opposition to this plan to the Turkish government (many members of which are themselves appalled by the actions taken by the minister and the Highways Division). We would like you to assist us in making the government realise that cultural and historic preservation is not only undertaken when an area in question has no other use, but that it is a real and important concept in contemporary life. We would, in short, like you to help us in whatever ways you can to make the preservation of this site of cultural heritage a reality in Istanbul … We are determined to oppose this plan by whatever legal means available to us, for ourselves and for future generations, who would otherwise be cut off from even a glimpse of the historical significance and beauty of their city, and indeed their country. Many diverse groups of citizens have already banded together to voice their concern and discuss strategies to stop this project. We believe that international support is crucial in this endeavour (www.arnavutkoy.org).

This piece calls for help from anyone who shares the same ideals with the initiative. These eloquently and emotionally ('the need to believe in paradise') expressed ideals focus on the preservation of both culture and environment:

> The construction of this bridge, the planning of which has been done behind closed doors, would destroy one of Istanbul's last remaining residential sites where historical Ottoman and even pre-Ottoman architecture exists as part of the daily lives of a wide variety of people. As evidence of the broad-based recognition of its importance, central as well as municipal governments have, for many years, applied strict regulations to construction and renovation of any kind in Arnavutköy. Arnavutköy itself is considered a 'historical site'. All buildings are bound by strict regulations to maintain all original facades, down to the details of the window frames (www.arnavutkoy.org).

With these political (and emotional) pleas, the ASG campaign creates its public and politically conscious profile through its formal environmental agenda, printed in press releases and published on the initiative's website which focuses on the protection and preservation of Arnavutköy. The ASG's environmental consciousness is conspicuous also in everyday discussions with the inhabitants of Arnavutköy. Many of my informants – both in interviews and in everyday conversations – claimed that one of the main reasons behind the objections to the bridge was the thought of what it would do to the natural and cultural life of Arnavutköy. One thirty-four-year-old woman of the group of less active participants stressed the following during one of our afternoon walks by the sea: 'There are many reasons that justify my choice to be against that bridge, but the most significant of them all is the destruction of the environment.' 'Look at this place,' said one of the non-active informants, pointing at the Bosphorus during a coffee pause in one of Arnavutköy's sea-front coffee-houses. 'How can anyone even think about destroying this natural beauty by placing a huge concrete bridge over it?'

Comments such as these suggest that the ASG is definitely a group with environmental references. The calls for environmental protection and preservation of Arnavutköy in the initiative's website and in press releases that constitute the formal presentation of ASG and the informal accounts of the participants mentioned above illustrate that ASG is indeed an environmental group. However, in addition to its environmental – in terms of science – face, ASG has distinct cultural characteristics which means, as will be discussed below, that the connotations that environment takes are as different as the cultural context in which it is perceived.

Environmentalism as a cultural characteristic

In one of her visits to my house, an active participant friend of mine brought photographs of past ASG festivals. One photo included a bulletin board with a picture of Atatürk – whose figure alone annuls any opposition whatsoever – accompanied with the comments 'The first and the most important environmentalist' (*İlk ve en önemli çevreci*). My friend noticed that I paid more attention to this particular photo than the others and she asked me if I liked it. 'It is an interesting picture,' I replied and she said: 'Atatürk was ahead of his age. We speak about the significance of the environment today when he [already] spoke about it many decades ago.' This incident suggests that even though the environment is often used to strengthen the arguments against the bridge, it is the personality of Atatürk which legitimates the environmental cause. Using dominant and popular figures along with the appropriation of their discourse is a common tactic for environmental groups which want to make their objections (see Chapter 5). However, this does not mean that the use of such symbols is purely instrumental. The previous account, as well as the next one suggests that the way in which the environment is presented is informed by the cultural character of the ASG; in other words, environmentalism is appropriated according to the cultural values of ASG participants. The following excerpt from an interview with a sixty-year-old less active participant is an example of the centrality of these values in contemporary Arnavutköy life:

> When we think of environmental conservation we usually think of material things like the pollution of historical buildings, monuments etc. But the cultural environment, to my mind, is as important as the material environment. People are not aware of this, they are trying to [...] remember all those things. When you tell them they will say 'Ah, the good old days.' And when you ask them what was good about those old days they begin to tell you about things which I think are very important ... Arnavutköy people are reviving their traditions and they are encouraging people to start doing

this. Perhaps it won't be commercially successful like it was before but at least people will be able to say that in Arnavutköy the strawberry cultivation started again.

He goes onto say: 'Another point I find very interesting is during Easter. Arnavutköy was very famous for this. You know they put *retsina*. You know the white stuff [resin].' And emphasises:

> It smells very special. The whole Easter week, the whole Arnavutköy would smell of this special smell; *retsini* and the bread. There were a lot of bakeries and there was a tradition. At least one of the bakeries used to bake bread for the poor and it was distributed freely. They are trying to recreate some of these traditions as well. Bring back the memories of these days when people used to give, during the holidays and each others special festivals. I find this very important, is what I call cultural environment. It is not something material, it is the immaterial conservation of things lost from the way of life.

In this way, the environment becomes a symbol through which shared meanings of a much more general social character, such as charity, are being expressed and promoted. These two examples described above indicate that Arnavutköy District Initiative is not merely an environmental group. Its actions suggest that it should be seen as a group with multiple faces. Even though environmentalism is clearly a central guiding principle, ASG revolves around many more issues that transcend the environment.

ASG as a political–transenvironmental mobilisation

As my fieldwork revealed, many of my informants were deeply concerned about environmental degradation beyond the community of Arnavutköy. This was demonstrated by, for example, the fact that many of them were purposely not car owners and those who were, drove only when necessary. Other indications of their general sensitivity to the environment included washing dishes by hand (rather than in automatic dishwashers) and methods of rubbish disposal. It took me a couple of months of living with my housemate to discover that she had her own car. When I asked here why she never used it, she replied: 'I don't really need a car in Arnavutköy. I go everywhere on foot and when I want to go somewhere downtown I use the bus or a taxi. I even think of selling it. I don't need to show off. I need to protect the environment and my neighbourhood.'

Furthermore, the ideals behind the initiative do not stop at environmental and cultural heritage preservation. ASG participants also appeal to ideals of democracy, as emphasised in an open letter published on the Fellowship of Mega Revma website: 'As citizens, we are very concerned with the blatant disregard of the democratic process. As residents, we are horrified that such an act of destruction could even be contemplated, and we are particularly horrified that this could come from politicians and technocrats of the central government with no stake in the city and its residents' (www.megarevma.net).

This comment of one of my informants combines environmental awareness with thoughts on progress:

> The atmosphere now is such that the prime minister and his government want to enter the EU. 'We are civilised ... we want to be European.' If you want to be European, the last thing you would do is to destroy the environment. Because this is not done, Europe doesn't do this. So, if you really, genuinely want to be part of Europe, you cannot do this (LAP, sixty-year-old architect).

This position merges with ideals concerning human rights, for example in the slogan 'Our neighbourhood will not be destroyed' (*Semtimiz feda edilemez*) which appears in the home page of the ASG website.

Along with environmental, democracy and human rights claims, the ASG demands certain policy-making. For example, whenever a bureaucrat announces the intention for the construction of the bridge or a media columnist supports the bridge project,[1] ASG responds with scientifically informed press releases, proposing alternatives to the bridge such as the Marmaray tunnel and the increase of the existing swift ferry boats that cross

1 It has to be mentioned that the cases of newspaper columnists who support the construction of the bridge are very few.

the Bosphorus. To achieve results, ASG participants have been in touch with politicians who have stated their support and they have also been in contact with the Ministry of Transportation to state their disapproval of the project and to discuss their objections. On 27 June 2002 a delegation of ASG participants visited the Turkish Parliament, met with deputies and scientists, and officially declared their opposition to the construction of the Third Bridge.

Throughout the course of its existence, the ASG has been supported by opposition political parties. Even though the initiative is not affiliated with any political party, participants are in touch with the local authorities, such as the *muhtarlıks* of both Arnavutköy and Kandilli, and have invited them to support their protest. They are in communication with the Municipality of Beşiktaş, whose representatives occasionally attend ASG events such as the festival, and with the Greater Municipality of Istanbul, mainly via press releases. In this way, the protest implies that it is not just about the bridge but also about the practice of democracy and the action of responsible citizens who demand their place in the decision-making process.

The political parties seem to have accepted the AGS's claim to being a robust part of the decision-making process in issues beyond the Third Bridge; e.g. wider environmental concerns. This acceptance was particular evident during the 2004 local pre-election period during which (a few weeks before the elections) candidates toured their electoral regions in open double-decker buses decorated with their party's flag. The loud music played on board was regularly interrupted so that the candidate(s) could address the central issues of their campaign platform. When the buses entered Arnavutköy, the candidates specifically addressed the issues of the pollution of Bosphorus and the traffic problems of Beşiktaş, proclaimed their opposition to the construction of the bridge and promised public dialogue on the issue. The protest appears to have acquired political status in local politics, thus it has become visible to the wider public beyond the confines of a neighbourhood political action.

These events suggest that the ASG goes beyond its initial environmental cause. For ASG participants the environment is a political issue and needs to be addressed as such. The articulation of environmental with wider social demands such as democratic participation and human rights as well as the appeal for politicians, NGOs, professional chambers, and other individuals and groups to take a stand, indicate that the ASG tries to address a wider range of issues and mobilise larger society, beyond Arnavutköy. According to these attributes, the ASG is not just an environmental group but a transenvironmental one as well.

The ASG as a community organisation

The ASG's transenvironmental aspect stems from the functions of the initiative as a community organisation. The annual festival, dinner parties and grand dances help raise community awareness and strengthen the ties of solidarity not only between ASG participants but also among the residents of Arnavutköy. The words of Ayşe (who will be discussed in the next section) and the opinions of a less active ASG participant are characteristic of the ties created:

> I feel very fortunate of being part of this group of people. If I hadn't participated in ASG, I would never have met all those wonderful persons I met with whom I feel so close. I would never have found out how much in common I had with my neighbourhoods. Some of them are like my soulmates (Ayşe).

> They say (ministry officials) that 10 million people are more important than 5,000 people but where will this stop? Today 5,000 people are not important, tomorrow there will be another issue and they will say 1 million people are not important. As far as I am concerned even one person is very important. One family is very important. The motorcar destroys something very valuable which is more than 1,500 years old. A stupid machine which is going to disappear anyhow! That machine is destroying not only my house but the way I live with my neighbours. It's destroying my butcher, it's destroying my barman and I have a way of life, my way of life is being destroyed and I hate living in apartments. Apartment living is a very unattractive way of living. A lot of people who live in apartments they don't even know their neighbours next door. It's terrible! And this way of living in a neighbourhood the Turkish *mahalle* is so attractive and that is why people are coming and try to buy houses or to rent in a place like Arnavutköy. Because this is a better way of life, for children, for families, for everyone (LAP, sixty-year-old architect).

As established in the introductory chapter of this book, the ASG was formed in 1998 (Danışman and Üstün 2000a, 2000b, 2003), an event followed with provisions for weekly meetings (usually every Tuesday evening), which take place on a regular basis to the present. The meetings are attended by a core group of regular participants as well as by several people who are brought in as resource people, for their expertise and connections. In the meetings, participants discuss any new development on the issue and tasks are delegated; for example, if a press release is considered necessary, then the spokesperson of the group is charged with the task. Every new press publication and press release is photocopied and archived; in fact, the ASG archive contains almost everything ever written on the subject. Those in charge with the Internet website (www.arnavutkoy.org) update the page whenever it is considered necessary and send e-mail messages informing all subscribers to the e-mail list of ASG.

The e-mail list (arnavutkoysemtgirisimi@yahoogroups.com) also functions as a communication site for issues that do not concern the Third Bridge. As well as sending e-mails concerning the developments on the Third Bridge issue, the members of the e-mail list communicate with each other, sharing information on cultural and social events such as films, concerts, demonstrations, press conferences, dinner parties, cultural and architectural studies, various contests (e.g., photography competitions), and public service announcements about missing pets and houses for rent. In addition, the list is used for circulating comments on political causes such as the Armenian issue, greetings during Christmas and other holidays, e-mail campaigns for signatures on petitions supporting human rights and other social issues, e-mails for help on personal dramas and internet technology. They also use the list to contact similar associations in other neighbourhoods and distributing information about the ASG annual festival.

The Annual ASG Festival (Panayır)

Ever since 1999, the ASG festival has taken place in Arnavutköy in the first weekend of June. The festival includes small bazaars, speeches, concerts and other artistic performances such as picture and painting exhibitions.

The preparations of the festival start as early as April to make sure that everything will be ready by June. The organising committee meets almost every week and discusses which events will take place, who will be the co-ordinator of each event and how they will be organised.

The *panayır* is set up in the central square of Arnavutköy and it begins early in the morning with speeches on the Third Bridge issue while, in the meantime, kiosks with food and other items welcome the visitors. Banners and placards with anti-bridge slogans surround the square and informative material is also available at kiosks. People come and go the whole day strolling around and watching the picture and painting exhibitions until the end of the festival in the evening when concerts by famous (and lesser known) artists are held.

From anthropological perspectives, festivals are examined as events which bind participants into wider frameworks of social action. For example, Clifford Geertz (1993: 452) claims that 'mass festivals assert the spiritual unity of the participants against their status inequality and project a mood of amity and trust'. Similarly, Esman (1982), following Durkheimian logic, examines festivals as mechanisms which reinforce social order. Jane Cowan (1990) examines public celebrations and festivals as mechanisms which symbolically express public and shared meanings. From a different perspective, Lewellen (2003) argues that ceremonies encompass political qualities and by studying them an anthropologist can trace the location of power and its manifestations.

All of the above dimensions are exemplified in the ASG *panayır*. It is a symbolic way of asserting the unity of the participants in their social struggle. The preparation alone is a process that unifies those involved in it. Aside from the organisers other people are involved as well. Arnavutköy citizens want to contribute somehow to the event; for example, many residents offer old objects or objects they don't use to ASG to be sold during the festival at a symbolic price. In this way, both buyers and sellers satisfy the sense of generosity and charity, both intrinsic features of participating in a group. The event is also supported by activists who are sympathetic with the ASG cause and find the festival as a chance to state that support and express wider issues and politicians who pass by and declare their support or listen to ASG's positions.

During a break between songs, the performer shouted: 'Arnavutköy needs many things, the Third Bridge is not one of them.' A friend and ASG participant wanting to stress the significance of the festival commented to me, 'Is this a revolution or what?' Hence, the festival appears to be an arena in which the political aspects of the conflict over the Third Bridge are expressed in various ways. For the participants of ASG, the *panayır* is more than a celebration; it is an open action of protest to the various 'faces of the state'.[2] For example, the figure of Atatürk (see previous section) symbolises the secular beliefs of the ASG participants. Therefore, the presence of Atatürk photographs in the festival is a metaphor for a struggle not only against the state that wants to have the bridge built but also against the state that is non-kemalist and non-secular. Another example of the political aspects of the struggle expressed in the festival is in the content of the various banners and placards, which, in addition to expressions of objections to the bridge, include written demands about policy decisions such as the construction of the tube tunnel or the creation of a fire department in Arnavutköy for the protection of the timber houses. 'We call for your support. Not just for the Third Bridge but also for our efforts to found a local fire-stations for our neighbourhood,' said a participant in her speech during the festival.

Various other activities, with similar role with that of the festival, are being organised. These activities include dinner and tea parties, usually with speakers, as well as awareness meetings about issues concerning the environment, especially the damage that the Bosphorus suffers from tanker crossings and the air pollution caused by the traffic on the two bridges. In these meetings, papers concerning the history of the area are also presented resulting from the launching of two research projects. The first project is an oral history project (Danışman and Üstün 2000a, 2000b) and the second is about a civil dialogue between Greece and Turkey, in collaboration with the *Fellowship of Mega Revma-Taxiarchis*, a Greek association of people who left Arnavutköy mainly during the period 1955–74, and which also supports

2 This expression is borrowed by the title of the book of Navaro-Yashin *Faces of the State: Secularism and Public Life in Turkey* (2002).

the anti-bridge campaign. According the volunteers of this project it aims at making Arnavutköy's history available to the wider public.[3]

During the course of my fieldwork, I had the chance to attend one of these gatherings. The grand dance was a formal event; invitations were printed and distributed and it took place to a restaurant which was booked well in advance exclusively for the ball. The tables were numbered and reservations were booked for specific people. However, after all the guests arrived and the dinner was finished, some guests, especially ASG participants, circulated among the tables, chatting with the guests and often sat down at other tables, drinking and having fun with the occupants. In this way, many of the people who joined the ball had the chance to hear the social news of the neighbourhood. For example, my housemate commented, 'I just found out that a friend of mine is getting married.' More importantly, many of the guests, most of whom were non-active ASG participants, had the chance to find out all the latest developments on the bridge issue. Additionally, during the ball apart from the entertainment, during a music intermission, ASG active participants presented update about the work in progress of the initiative and the upcoming festival.

The above events as well as the e-mail list constitute a locus with various and multiple functions. All these events bring together the neighbourhood and reassert the commitment of the participants to their struggle at the same time they express their opposition and stance to the outsiders. The e-mails circulated in the ASG mailing list, represent the life mode of a community with various, often heterogeneous interests, which nevertheless accepts and participates in the struggle against the Third Bridge and through e-mails suggests how an ideal community should be. The festival and the other events, operate in support of local solidarity at the same time

3 Another ASG project is the establishment of a local fire department. The people in charge try to get public funding in order for the department to be established, based on the argument that since there is a large number of timber houses in the area firemen forces should be at hand. Finally, ASG participants are actively involved in a campaign against the reduction of the ferries that cross the Bosphorus Strait between the Asian and the European side named '*Vapurumu Vermiyorum*' (I am not giving up my boat) (www.vapurumuvermiyorum.org).

that the participants and those who attend these events are informed of the new developments of the Third Bridge issue. In addition, such events constitute a chance for ASG to find out who supports its cause and to what extent, and to disseminate information and symbols of the struggle. This brings us to the next section, which discusses the association of the ASG with the wider society.

ASG as an urban protest group

ASG's activities make it as much an urban (Castells 1983; van Naerssen 1989; Pickvance 2000; Mayer 2000; Hamel 2000) as an environmental group. This is clear from the views and demands expressed by ASG participants and below I refer to a conversation I had with a LAP, a forty-year-old psychologist:

> I am convinced it will be a disaster, anywhere, not only here. A Third Bridge will destroy Istanbul and its traffic because it is not an alternative. It will make things more complicated. There will be pollution, it will destroy the history, the culture of the area and anywhere else in Istanbul. We already have two bridges and we are not solving any problem. Public transportation is the only way. Building a bridge means the car sellers the petroleum industry will get richer but the people will not benefit.

She went on to say:

> And it is really the stupidest thing. Generally there is a plan about Bosphorus and what this plan says is that people should approach Bosphorus from behind the hills. They should park their cars in parking spaces up the hills and then they should come down by public transportation. There should be frequent transportation between the Etiler, for example, Bebek and Arnavutköy. And then people should use the sea ferries and they should be free of this traffic. At Arnavutköy, you know this double road goes until Bebek and then to come to our University there is a single road which is the stupidest thing. You spend a lot of money, build these pylons into the sea, make this double highway and then it stops after Bebek. *Yani* [that is], it's really stupid! Because of this street and noise the Yalis in Arvavutköy have suffered from pollution and people don't live there any more. They have turned them into business places.

> But those people used to live in front of the water! Now they have this huge noisy street! So noise pollution is another major problem. And of course traffic accidents, speeding cars is detrimental to the public. The people have been cut off from the sea! The sea belonged to them! Now is so difficult to negotiate this huge highway.

Such views were developed through the course of ASG's development dating to December 1998 when meetings started to take place and the residents began to think and plan an anti-bridge action. According to ASG sources, the Marmaray Project (see Chapter 2), along with the improvement of the existing public transportation system of Istanbul, is the most viable solution to the city's traffic congestion problems. As a good number of my informants clearly stated, Arnavutköylites do not want a bridge to be built in their neighbourhood and they are ready to prevent it from being built. What has been an undisputable certainty among them is that behind the bridge there were interests forcing the Ministry's authorities to construct the bridge. Furthermore, my informants' arguments concerned the negative effects that the construction of the bridge would have on the natural environment and on the built cultural assets of the area.

> When I first read the news about the bridge I felt like a thief was breaking into my house. From the very beginning, I decided that I would never let it happen. [Her argument continued] Some people living under the second bridge did not feel the earthquake[4] because their houses were shaking all the time due to the traffic on the bridge! (AP, forty-four-year-old high-school teacher).

> Many people were against it from the first moment, we started discussing it and just like that, ASG was born (AP, sixty-year-old pensioner).

> Academics and other experts informed us about the harmful effects that the bridge will have on our surroundings (AP, fifty-year-old businessman).

> Do you know how much noise and air pollution this bridge is going to cause? Why don't you take a walk under the other two bridges? (AP, thirty-three-year-old preliminary school teacher).

4 This person refers to the big 1999 earthquake that hit the Marmara region of Turkey, and Istanbul in particular, and resulted in the deaths of 14,000 people.

By the beginning of 1999, the conflict escalated between those who were for and against the construction of the bridge. During that year, mainly in the national but also in the international press, about 120 articles and press releases were dedicated to what came to be widely known as the Third Bridge (*3. köprü*). The conflicting parties were the residents of Arnavutköy represented by the ASG and the Ministry of Public Works and Settlements represented by the 17th State Highways Department (*17. Karayolları Bölge Müdürlüğü*). In addition, ASG lobbied with similar initiatives in other countries and with international NGOs. As an AP stated:

> ICOMOS has listed Arnavutköy as HERITAGE AT RISK. So there is International support. We have been in touch with a lot of European Universities. For example, in Barcelona there is an organisation of Mediterranean Cities. They are waiting to hear from us. If we say to them 'Arnavutköy is in danger' they will start sending e-mails, they will protest etc. so I think that this will make the government think twice, or three times before taking any stupid action (LAP, fifty-year-old architect).

In this way, ASG tries to 'render the struggle against the Third Bridge a common cause and thus achieve the status of a social movement by having its objectives undertaken collectively by the mobilisation of a distinct social base and by rendering its activity towards changes in policy direction' (see Lowe 1986: 3). The launch of the ASG website (www.arnvutkoysemtgirisimi.tr) can also be seen as an example of the effort of ASG to communicate with the rest of society both in national and international level. At the same website, ASG hosts opinions of traffic experts, a fact which conforms with Castells' opinion about the connection between urban mobilisations and professionals.

In one of the most acclaimed works in the literature on urban movements entitled *The City and the Grassroots*, Manuel Castells (1983: xiii) gives a detailed analysis of the characteristics of urban movements, and argues that there are four basic elements which define them. 'The first is that urban movements articulate the three goals of collective consumption demands, community culture, and political self management. The second is that they are self-aware, and the third is that they are connected to society through media, professionals and political parties.' Nonetheless – and this is the fourth element – although connected to the political system, the urban movements that he is describing are autonomous of any political party. For Castells, urban protest movements in our societies and in our epoch particularly seem to be developing around three major themes: demands focusing on collective consumption, in terms of goods and services provided by the state; defence of cultural identity concerning a specific territory; and finally, political mobilisations regarding the state and particularly the local government. Like Castells, van Naerssen (1989) defines the main goal of urban social movements as the improvement of individual and collective consumption within marginalised local spaces, while Schuurman (1989) argues that they are social organisations with a territorially based identity striving for emancipation via collective action. For Pickvance (2000) an urban movement is conceived as an individual organisation which makes urban demands in every level. In a slightly different direction, Mayer (2000) offers a definition that contains other parameters. In his view, urban movements strive toward a more participatory and sustainable city by challenging its shape and its regulation as well as the undemocratic and un-ecological urban development. Similarly, Hamel (2000) claims that urban social movements seek to redefine local democracy. They challenge traditional forms of local politics by searching for new meanings responding to the new global order. They focus on urban development in terms of state intervention and the re-definition of it. They promote local democratic processes and suggest alternative models of economic development, contradicting the dominant model. Finally, they advocate a developmental model that is not based on principles defined by market priorities.

ASG's actions such as the press releases and the association with traffic experts and other scientists, suggest its connection with wider circles in Turkish society, such as the wider population of Istanbul and other similar initiatives, a fact which also agrees with Castells when he claims that urban movements aim at being connected to society through the media and professionals. Similarly, the publicity that the initiative has gained through the media and the discussion initiated by this publicity gives the chance to communicate not only with the rest of society but to implicitly suggest changes in policy. The role of the media in this effort, as will be seen below has been significant.

The role of the media in popularising the ASG protest

By the end of 1998, ASG had press releases locally and nationally. Those first press releases stated the opposition of ASG to the bridge project. Some of the titles were the following: 'We do not want a bridge in the neighbourhood' (*Mahallede Köprü İstemiyoruz*), 'Meeting for the dispute over the Third Bridge' (*3. Köprü Tartışmaları Üzerine Toplantı*) and 'Why do they want to build the Third Bridge?' (*3. Köprü Neden Yapmak İstiyorlar?*). As mentioned above, the Third Bridge protest has been picked up by national and international Media. Ever since 1998, there have been hundreds of newspaper articles on the Third Bridge issue, most of them in the national press. Since 1998 hundreds of articles in almost every mainstream political newspaper have been discussing the issue from various viewpoints questioning the necessity of a Third Bridge.

On 20 September 1997 an article about the need for a third Bosphorus bridge and the opinions by scientists and lay people about it was published in *Sabah* newspaper. In the article a Professor at the University of Bosphorus (Boğaziçi Üniversitesi) argued that the solution for the traffic congestion that Istanbulites experienced was not a third crossing but an improvement of public transportation. In addition, the newspaper mentioned a survey conducted by a survey company (Parametre Araştırma) in which 81 per cent of the informants answered that if a third crossing was chosen as a solution they would prefer an underwater tube tunnel and only 17 per cent stated that they would prefer a Third Bridge. In addition around 60 per cent of the sample believed that a potential construction of a bridge would damage Istanbul's environment (*Sabah* 1997).

The author of an article from the same newspaper, published on 18 June 2002, wonders whether a Third Bridge between the two existing is necessary: 'The two bridges are close enough. A third one between them is not a good idea [...] After all, a woman puts on only one or two necklaces maximum. A third one is an exaggeration' (*Sabah* 2002). Another article in *Akşam* entitled 'Does Istanbul need a Third Bridge?' discusses various opinions regarding the usefulness of a third Bosphorus crossing (*Akşam* 2002). Other articles focus on the various interest groups and discuss who and why want the construction of the bridge. In an article from the newspaper *Cumhuriyet* entitled 'Who wants a Third Bridge over Bosphorus', the author claims that even if ten bridges would be built, they would never be sufficient for a country whose mentality is being shaped by TV commercials that urge consumers to buy new cars. The article concludes with the question 'Is this our country? ... The people do not want the Third Bridge. We do not want it' (*Cumhuriyet* 14 February 1999).

Secondly, they focus on the usefulness of the underwater tunnel and whether or not its construction will render the construction of the Third Bridge necessary. On 7 June 2003 the author of an article published in the newspaper *Radikal* wondered whether the Marmaray Project will be done. An article in *Sabah* claimed that even the Minister of Transportation has the wrong idea about the tube tunnel project, since he claimed that the number of cars that will cross the tunnel will increase to 70,000, while the truth is that the tunnel will accommodate only railway vehicles (*Sabah* 26 January 1999). In an article in *Cumhuriyet* traffic experts discussed the possible scenarios that would ease traffic congestion (*Cumhuriyet* 5 May 1999) whereas an article in *Hürriyet* referred to the conclusions of a panel held at the Marmara Hotel concerning the usefulness of a third crossing over Bosphorus (*Hürriyet* 4 May 1999).

Thirdly, the majority of the articles focus on the ASG and its opposition to the government. For, example many newspapers have written about the ASG festival in articles titled 'A festival protest for the Third Bridge' (*Cumhuriyet* 1 February 1999), 'Environmental Coalition' (*Milliyet* 1 February 1999), 'Do not touch our Bosphorus' (*Milliyet* 11 June 2001). The English speaking newspaper *Turkish Daily News*, in an article published on 14 June 2002, refers to interviews taken by Arnavutköy residents regarding their opinion about the construction of the bridge: 'We will continue struggling to protect Arnavutköy which has a history of 1500 years'; 'We will put every effort to protect our village'; 'There are very old buildings here. Some of them will disappear with the bridge'. The Akşam newspaper refers to a demonstration organised by ASG on World Environment Day, characterising it as an environmental protest (*Akşam* 7 June 1999). A similar article was published in the newspaper *Star* on the same date (*Star*

7 June 1999). The newspaper *Yeniyüzyıl* described the ASG struggle in two articles (*Yeniyüzyıl* 15 and 22 May 1999).

The foreign press places the ASG struggle as central in the Third Bridge issue, with items appearing in the *Los Angeles Times*, *Washington Post*, *Le Monde* and *National Geographic*. An article appearing in the *New York Times International* has already been discussed. The message of another international press article is examined here, from the *Los Angeles Times*. The article, titled 'A Bridge that Turks Refuse to Cross', focuses on the struggle emphasising the capacities of an organised resistance, an element lacking in previous years in Turkey, especially since the 1980s. 'A generation after the army overthrew a civilian government and crushed street fighting between right and left-wing students, at a cost of thousands of lives, Turks are still reluctant to join in any organised challenge to the state. Those who do, be they Islamists seeking the right to wear head scarves in public buildings or Kurds demanding schooling in their own language, usually face arrest.' The *Washington Post*, in an article entitled 'Historic Homes Torn By Times in Istanbul', refers to the ASG resistance as part of the renovation efforts of the residents of the historic old houses of the Bosphorus. The article reveals the author's surprise that a civil protest of any magnitude might exist 'in a country where public activism was largely quashed by a restrictive constitution imposed by the military in the 1980s' (*Washington Post* 2001).

All the above themes have been repeated since 1998 in many newspapers. Most of the articles focus on the formation, the arguments and the activities of the ASG. Therefore, the role of the media has been very significant in the case of the third bridge, in terms of making the issue public and by creating a communicative space between citizens on wider political issues. In this way, the ASG opposition has generated a discourse which involves much larger issues than the local discontents agitate. Debates about whether a Third Bridge is needed and the projected effects of its construction, as well as the comparison of the bridge to the underwater tunnel, result in a more encompassing discussion about public and private transportation. In turn, the discourse which develops concerns policy decisions and the role of the welfare state in an open market society.

These issues are constantly put forward by the ASG and its participants. The evidence shows that ASG is a multifaceted group, as much of an environmental group as a transenvironmental one. This means that for ASG the protest against the Third Bridge involves many issues and this has shaped its character from a mere environmental group to a community organisation and an urban protest. However, the definition of a group as urban in turn, involves a wide range of factors that should be taken under consideration. These are: individual socio-economic factors, ecology, life experience, informality of urban organisations and degree of integration between individual urban situations (Roberts in Schuurman 1989). Ho Kwok-Leung (2000) claims that in order to understand urban movements we must take into account both the actors and the social and political contexts. Pickvance (2000) argues that the analysis of urban movements should include urbanisation processes, the role of the state, the development of the middle class and the general economic and social conditions. Following these, I believe that if we wish to have a better idea of the ASG we should, among other matters, have a look at the identities of its participants, which is the focus of the next section.

ASG: Arnavutköy Semt Girişimi

ASG is an informal group of fifteen people who live in a middle-class neighbourhood of a global city. This group, supported by the majority of the residents of the neighbourhood, was christened the Arnavutköy District Initiative in 1998, and formed in order to protest against the proclamation for the construction of a bridge above Arnavutköy. Since this description tends to characterise the ASG as merely an interest group the actions of which are being triggered by NIMBY concerns, I suggest examining the identities of its participants as one of the main reasons that sustain their struggle. The reason I chose this analysis stems from the belief that ASG is much more than a local NIMBY protest. On the contrary it is a collective action with political references. As Jamison et al. (1990) claim, contemporary social movements combine identity and political action. Therefore, in order to examine such movements we have to focus upon the actual strategies of environmental organisations but also the examination of the identity formation of their participants. This approach allows for an analysis of contemporary social movements both as the quest for new socio-political identities and as political activity aiming toward achieving certain tactical results.

Before discussing the composition of the group I should first mention that the ASG is not an official organisation in terms of having a statute and thus an official recognition by the state. Because of this, I refer to the people who participate in ASG activities as 'participants' and not as 'members' since there is no subscription needed to be involved in its activities. According to my informants, the decision not to ask for official recognition was deliberate. As one of them told me, if they had claimed official status, the bureaucratic obstacles would have been most likely impenetrable. The official formation of organisations that have political aspects in their agendas – such as ASG that criticises the way environmental and traffic policy is implemented – tend to be rejected by the state. This is probably a remnant of the 1982 Constitution, which weakened the freedom of movements, associations, and demonstrations (Beşpınar-Ekici and Gökalp 2006), even though the Turkish constitution as amended in 2001 mentions that everyone has the right to form associations provided that it does not threaten national security, public order, morals, health or protect crime (The Constitution of the Republic of Turkey, 2001). Official status is granted to the Bosphorus Association of Arnavutköy (*Boğaziçi Arnvutköyluller Derneği* – BAD), which is a local cultural association and the participants of which are ASG members. Nevertheless, it should be noted that all ASG participants are members of BAD, but not all BAD members are participants of ASG.

When I was conducting my fieldwork, the main core of ASG consisted of nine persons who were always present in meetings, organised the events and were always informed about all the new developments of the Third Bridge issue. Most of them were involved with the campaign since the beginning of the struggle. I refer to these persons on as 'active participants'. Five persons were less involved but they still felt very strong about the issue and even though they would attend meetings and other events less often than the active participants, they were always informed about any new developments of the group and the Third Bridge issue. I refer to these people as the 'less active participants'. Many people have been involved with ASG, but not in a systematic way. Some of them were active in the past but at the time of my research they were not involved at all, others take part in meetings and events occasionally, and some of them were occasional affiliates since the beginning. I refer to this group of people as 'non-active participants'. Finally, there are the supporters who are not residents of the area but have participated in meetings and other events and have contributed to the campaign. Most of them are scientists and/or experts on traffic and architecture subjects; others are activists and there are also few celebrities who have publicly stated their support. These people are what Schönwälder calls 'power holders' who have the power to ask for crucial measures to solve the problem (Schönwälder 2002).

In this chapter I use the following definition of activists:

> Social movements include not only activists but also supporters. Movement activists are those who are committed to public actions intended to influence the behaviour of the policy system and of the broader population. Committed activists are the core of a movement and have been the subject of much recent work in the social movements literature (Stern et al. 1999).

For Stern et al., the movement becomes an important part of their life and a central element in their identity. Movement supporters are those who are sympathetic to the movement and who are willing to take some action and bear some costs in order support the movement. Of course, the boundary between supporters and activists is fuzzy and people often move back and forth, being activists for a time then retreating to a less committed but still supportive role.

Following this definition, I would suggest that in order to understand the role of the active participants, one should examine them through their leftist and secularist identity – on which I will focus later in this chapter – in relation to their demographic characteristics. To these individuals, the bridge project is not an isolated issue. It should be seen as a result of decades of liberal economic politics that 'serve the interests of the few instead the interests of the people', as one them claimed.

The role of the less active participants of ASG is more clear when seen under the prism of their global citizenship. This sub-group of ASG seems sensitive to issues existing in the national and international agenda of civil society. For them, the Third Bridge is one of these issues, seemingly dissimilar but part of the same discourse, namely that of the global civil society. The non-active participants are those who do not identify themselves with a greater community. For them, Arnavutköy is their identity, their place.

The supporters of the ASG are very close to what Beşpınar-Ekici and Gökalp (2006) call the 'power brokers'. According to their definition, the power brokers of CSOs are middle-class professionals and academic circles with 'formal' expertise who assume the representation of the 'local'. In our case, they are scientists, activists, members of professional associations and celebrities. People who, due to their status, are entitled to articulate claims at the same level that the officials of the Ministry of Public Works operate, rendering ASG a dialogue partner with more powerful centres.

The identities of ASG

Is ASG a NIMBY protest?

In 1999, the *New York Times* published an article regarding the Third Bridge issue titled 'A New Bridge? Not in My Backyard!' The article reported that the plans for a new bridge across Bosphorus had set off a protest similar to the battles over urban renewal that broke out in many American cities during the 1960s and the 1970s. It also referred to interviews with Arnavutköy residents and other opponents of the bridge project, who underlined that a Third Bridge will not ease traffic and that Arnavutköy is a unique area that should not be spoiled by a modern construction such as a suspension bridge.

Mayer (2000) writes that middle-class opposition movements often achieve their goal to stop an undesirable development, and this in turn results in the reposition of the project to a working-class area. Even though the struggle against the construction of the bridge could be seen as NIMBYism, as suggested by the title of the article of the *New York Times*,

I would claim that it involves a lot more than that. As mentioned below, ASG participants do not only oppose the construction in Arnavutköy but in any other Bosphorus area. Besides, there had been objections to the consequences that the construction of the pyloned highway connecting Beşiktaş to Etiler would have on Arnavutköy's architectural character and social life long before ASG was established. Those objections were expressed by some people who are today in the heart of ASG, most of whom have been politically and socially sensitive individuals, as my acquaintance with them revealed.

As I was interviewing one of the people favourable to the bridge project, he mentioned that: 'Arnavutköylites would have reacted anyway. They do not really want to oppose the bridge. They want to oppose to everything because they are leftists (*solcu*) and they are using the bridge as an excuse.'

What the interviewee implied was that the opposition to the bridge was an alleged reason and not the real cause of the ASG campaign. That observation was valuable to my research because, even though the term leftist was negatively stressed in a manner that portrayed it, as Zygmunt Bauman (1986) would say, as the counterculture of modernity,[5] it helped me realise that the struggle against the bridge was not a monolithic case of NIMBYism. What my subsequent research revealed was that the campaign against the Third Bridge has been an implicit way to oppose to dominant policy making. 'We do not want to see a Third Bridge neither in Arnavutköy nor in any other district!' (*3. Boğaz Köprüsü' nü ne Arnavutköy'de ne de başka bir semtte görmek istiyoruz!*) (ASG placard) is one of ASG's slogans. When I asked one of the participants why they do not want a Third Bridge anywhere, he responded that it was not only that the environmental and cultural heritage could be damaged, but that a public statement would also be made: 'ASG would be heard.'

5 Michael Herzfeld (1991, 1995) makes also a similar point, when he discusses the debates over the preservation or not of the old town of Rethemnos, on the island of Crete in Greece. In his account, those who were against the preservation of the old buildings blamed the left-wing anti-business mentality of those who wished the old town to be protected and preserved.

Several studies suggest that local opposition often relates to lack of trust in the government, fear of health consequences and other ideological and cultural reasons (Drijver 1992; Theodossopoulos 1997, 2000, 2002; Bartu 1999; Haynes 1999; KHRP 1999, 2002; Mayer 2000; Mavridis 2003; Arsel 2003). Scholars such as Lober (1996) and Wolsink (1994) claim that in many cases, NIMBYism is not a useful tool to study and analyse public opposition. Kempton et al. (2005: 124) argue that the term 'NIMBY is generally used as a pejorative term, implying selfishness as an underlying cause. Secondly, it appears to incorrectly describe much local opposition; and thirdly the actual causes of opposition are obscured, not explained, by the label.' I agree with them and I claim that if a label should be put to ASG, then the NIABY label is more suitable. 'NIABY' is an acronym for the phrase Not In Any Back Yard and describes cases where people's opposition is based on common interests rather than self-interest. As Ho Kwok-Leung (2000) suggests, social movements should be seen as group and not individually embedded. In the case of ASG, such an examination should focus on the collective identities of its participants.

Active Participants (AP)

As mentioned above, the active-participant group of ASG consists of nine people. The reason that I place them in this group relates to their high level of involvement with the campaign. These people, even though almost all of them are professionally active, are involved with the campaign on an everyday basis. Doing something for ASG is part of their everyday schedule, as is going to work or running errands. Five of them are women and also five of them are university graduates (two undergraduates, one postgraduate, two PhDs) while the rest are high-school graduates. This fact, in a country where only 27 per cent of the population enrols at universities (http://www.turkeynow.org/db/Docs/Sector%20Reports/Sector%20Reports%202008/EDUCATION.pdf) and where the illiteracy rate is 13.5 per cent (www.turkstat.gov.tr) suggests that ASG active participants are very well-educated persons for the Turkish context. All of the active participants work except one who is a pensioner. Five of them are either self-employed or are freelancers and three of them are employed in education (preliminary, secondary and university). Only three of them are married and the rest are either single or divorced. My fieldwork indicates that their living standards are medium–high, if compared with the living standards of the majority of Turkish people. In this sense, I would say that they belong to the middle and upper middle class of the country as many of the residents of Arnavutköy and the other Bosphorus areas.

Many of the active participants pointed out the importance of politics in terms of the choices that the governments have made and the way policy making is carried out in Turkey. As one of them stressed:

> If ASG was formed in the 1970s it would not be so important. Back then everyone was into politics. After the 1980s this changed and collective action has not been part of our public life. Our society has become more individualistic and movements like ASG are rare; that is why it is so important (AP, forty-year-old tourist guide).

And when I conveyed to him the view that their campaign was accused of being a leftist reaction rather than an honest local mobilisation he told me that: 'Our initiative does not consist of leftists even though some of us do indeed have a leftist political orientation but you know something? Only a leftist can see things in a way that allow him to react and resist.' Less politically active informants also underline the importance of politics in the Third Bridge issue like the one below: 'It is interesting to see why different ministries have different agendas regarding the Third Bridge issue. It is all about interests' (AP, sixty-year-old pensioner).

New Social Movements theorists claim that social movements mobilise around common identities which, in order to be successful, should construct an enemy and move against it (Tarrow 1998). In the case of the active participants of ASG, the opposition takes the form of a leftist, secular identity against an Islamist, neo-liberal government. The following example of an active participant, illustrates this point.

Ayşe is a middle-aged woman born and raised in a big Turkish city. She is a college graduate married to a university professor, and the mother of two children. She could be described as a contemporary woman, with a nice house, a satisfactory income and a circle of friends of well-educated, upper-middle-class individuals. Her beliefs are clearly secular, a fact visible not only by her looks, as she does not wear a veil, but also by her arguments regarding the pro-Islamist ruling party. She made the following comment to one of her friends, in my presence, the day after the local elections of 2004, when the candidates favoured by AKP won almost every major municipality in Turkey. 'It is our fault that they won the elections. We shouldn't have let them gain all that power. Now we pay for our mistakes and they have taken over all the municipalities of the country.'

Her beliefs, however, are not limited to secularist ideals. During the *coup d'état* of the 1980s Ayşe and her husband had to leave Turkey due to their political beliefs. One day she told me:

> I like to think my self as a leftist and I am very proud of it. I don't care about mainstream ideologies and about all those who condemn lifestyles like mine and all those who claim that the bridge will benefit Istanbul, serving big corporate interests, even though they very well know it will not. I can sleep at night with no guilt and I don't care if I chose to wash my own dishes.

In fact, when I conveyed the opinion (mentioned earlier) of the supporter of the bridge project to her, similarly to the other AP mentioned above she said: 'We do not oppose to the bridge because we are leftists. In our initiative, there are individuals of diverse political persuasions. As a leftist, however, I believe that leftists are more sensitive than others to such issues.'

For the active participants of ASG, the bridge issue has not been an isolated problem to deal with. They see the decision to construct the bridge as one link of a whole chain of political decisions which has led back to a particular development model chosen by the Turkish state. Another one of the active participants said: 'The whole bridge issue started with the Marshall Plan that Menderes got. After that, roads were built and we [Turkey] started to become little America' (see also Keyder 1999). Similarly, the press representative of ASG who also belongs to the AP group of participants said:

> We don't just focus on the bridge. The bridge is just the cherry on a cake that started to be baked a long time ago. The technocratic mentality of the Turkish state started in the late 1940s when we accepted help from the Americans. It is when all started ... This kind of politics has been followed by all the governments. For example, Özal, who was prime minister in the 1980s, was known as the 'King of bridges'.

Along with references to leftist, the secularist identity came up very often in my discussions with most of the active participants. Almost all of them expressed explicitly their disapproval of the Erdoğan administration, and some of them were associated with secularist parties. Many of them told me that the best thing that ever happened to Turkey was the decision taken by the founder of the Turkish state to make the country a secular republic. Also, many of them stated their trust in the army as a safeguard of the secular establishment. One of my informants would hang pictures of Mustafa Kemal Atatürk in the front window of her house in celebration of Turkey's secular holidays (*bayramlar*); of course she refused to celebrate religious holidays in a similar way – the religious holidays, which, according to her opinion, were less important than the secular ones.

This book does not suggest that leftist and secularist identity lay behind the reason for the ASG mobilisation;[6] however, the evidence suggests these are the two main identities around which active participants unite and develop the solidarity needed for initiatives such as ASG. This solidarity fuses with the concerns of the rest of ASG, such as the environment and creates a platform for the expression of 'sameness' to the outsiders (Stephen 2005).

Less active participants (LAP)

During my fieldwork, five people comprised the less active participant group, three were employed full time: as a university professor, a newspaper journalist, and an English teacher. One was a housewife and one was unemployed. Like the active participants, most of them spent a significant part of their lives abroad, mainly for education. For them, the construction of the bridge represents a threat to their lifestyle and as a result, they protest against it. As Edelman (2001) informs us, protests born out of the threat on a group's lifestyle can be feminist, environmentalist, gay and lesbian,

6 It should not come as surprise that these two identities are intermingled since leftism, in terms of Marxism, is intrinsically linked with atheist and secular beliefs.

oppressed minorities movements that seek new collective identities. As it is evident from the next account, similar issues play their role in the formation of an ASG identity.

Deniz is transsexual. She has been living in Arnavutköy all her life. Her house is one of the houses that would have to be demolished if the bridge would be built. My first encounter with Deniz was during a BAD meeting which we both attended. What was surprising to the 'orientalist' part of my intellect was the acceptance of her 'preference' by the people who were also attending the meeting. Throughout my fieldwork, I did not notice any specific reactions to her, either in the ASG meetings or at different social occasions attended by Deniz. When I interviewed her regarding the Third Bridge, she said:

> I was born and raised here. I am both Arnavutköylite and transsexual. If the Third Bridge comes here, my struggle will start all over again. If the Third Bridge doesn't come, Arnavutköy for me will be my paradise. If the bridge comes, I will have to move and start struggling for my particularity. I definitely don't want the Third Bridge here.

Deniz implied that her effort to establish herself as an equal member of neighbourhood and participant of ASG had not been easy. Nevertheless, in her opinion she managed to establish a modus vivendi between herself and the rest of the community, something which the construction of the bridge would destroy. Therefore, Deniz's participation in ASG was mainly motivated by her own individual reasons, and her opinions about the bridge also express personal concerns. In Turkey, public statements about a taboo issue as delicate as transsexualism is the result of public demands made by transsexual individuals and transnational human rights movements concerning state recognition and respect (Kandiyoti 2002). For Deniz the demand for respect to be shown to her and her neighbours is much like the demand for respect to be shown to her and her transsexual friends.

Similarly, the profile of the other less active participants is not mainstream; for example, one of them is a single, middle-aged woman of non-Turkish origin who is politically conscious and socially active. An excerpt from an interview with her provides insight:

> Here in Turkey, in our country, people feel that they are guided and whatever the government does they accept it. They complain about it, they say it is good, they say it is bad but they accept it eventually. It is probably the first time in my life that I saw that a non-governmental organisation … I don't know if you can call it an organisation or a group, a very small group of people are struggling … at least they can say no, which is very good (LAP, forty-two-year-old psychologist).

Like Deniz and the woman above, those who participate in ASG have their own personal reasons for their involvement, but their motives stem from collective rather than private interests. These people consciously participate in the ASG initiative and previously they were active citizens in other communities. Their participation in those communities shapes their identities which are not merely individual self-definitions, but emerge from a 'civil society' discourse at national and transnational level.

Many times ASG participants state that they are part of civil society and that they are supported by NGOs and transnational organisations.

> We have been in touch with a lot of European universities. For example, there is an organisation of Mediterranean cities in Barcelona. They are waiting to hear from us. If we say to them 'Arnavutköy is in danger,' they will start sending e-mails, protests etc., so I think that this will make the government think twice or three times before taking a stupid action (LAP, sixty-year-old architect).

After the 1980s, the concept of civil society emerged in Turkey as a counterbalance to the statist influence. Activists who claim to be part of it declare that civil initiative can contribute to the stabilisation of democracy in the country and the same rhetoric is often used by ASG participants:

> The last five–ten years, civil initiatives in Turkey – Non Governmental Organisations – have become very important. Especially after Habitat II,[7] I think it was in

7 Habitat II was the name of the second United Nations Conference on Human Settlements held in order to address the issue of 'Adequate shelter for all' and 'Sustainable human settlements development in an urbanizing world'. According to the declaration of the conference, 'human beings are at the centre of concerns for sustainable development, including adequate shelter for all and sustainable human settlements, and they are entitled to a healthy and productive life in harmony with

> 1996 here in Istanbul, more than 10,000 people went. Habitat II was an initiator for a lot of NGOs, especially protecting women and children and environment. Many civil societies have been formed after that, and it has created an awareness of the importance of non-political, civil resistance, civil disobedience in this country. The governments are edgy too! They realize that this is a new force in the society. So, I think it will be very stupid for any government to disregard Arnavutköy and their resistance and say 'Who are these Arnavutköylites? I don't care, I want to build the bridge.' There will be a big problem and there will be an international crime. So, I think things have changed since the first bridge was built. The first bridge was built in 1973, thirty-one years ago. Turkey was a different country, Europe was a different place; now the man in the street, the civilian population is more important than before and that is, I suppose, part of becoming more democratic and more civilised (LAP, sixty-year-old architect).

European perspectives of Turkey and prospects for the country's future is another issue that appears quite often in my discussions with less active participants. Some of my informants stated that they wish to enter the EU as a nation because they believe that certain rights should be protected:

> The intention to enter Europe means that people have to reach certain standards of the protection of environment of the protection of civil rights, of the protection of democracy. So, it is definitely a positive thing. And it prevents bureaucrats, it prevents civil servants, governments from behaving very irresponsibly (LAP, sixty-year-old architect).

Less active participants of ASG have created a transnational identity by reflecting national and broader developments which – combined with their demographic characteristics – form an identity of global activists. And it is this identity more than anything else which motivates their participation in

---

nature.' The conference was held in Istanbul, 3–14 June 1996; it was attended by the representative delegates of 171 countries – a total of 16,400 people. The final document mentions that 'within a framework of goals and principals and commitments, a positive vision of sustainable human settlements where all have adequate shelter, a healthy and safe environment, basic services, and productive and freely chosen employment. The Habitat Agenda will guide all efforts to turn this vision into a reality' (http://www.un.org/conferences/habitat).

the ASG. In contrast, as will be discussed in the following section, the next group of ASG participants are motivated from more localist concerns.

Non-active participants (NAP)

The number of the non-active participants is larger than the other two sub-groups. In the course of my fieldwork I met dozens of people who belong in this category. Some of them I interviewed; others I did not; thus, I do not have a complete demographic profile for this category of participants. Even so, I would say that they are representative of all the types of people in the neighbourhood: working individuals, unemployed, students, pupils, housewives, pensioners, and so on. My fieldwork revealed that the majority of the residents of Arnavutköy are indeed against the bridge; in fact, no one told me that he/she supports the bridge – not in formal interviews or in discussions as part of everyday interactions. It is my overwhelming impression that their opposition stems from the fact that the construction of the bridge would detach them from their 'place'.

Tim Ingold (1993) suggests that local perception of place is revealed through experience from dwelling in a particular place. This relationship is better expressed by the concept of place as advanced by Tilley (1994: 34), a concept which 'privileges difference and singularity'. Knowledge of place depends on experience, and experience is translated into 'dwelling'. In this sense, place draws on the social practice of dwelling, the concept of which is epitomised by Macnaghten and Urry (2001), who define dwelling in terms of participation. For them 'human subjects are united with their environment and there is no distance between people and things' (ibid.: 6). In other words, dwelling is a social practice which describes a unified world between people and things. Similarly, Feld and Basso (1996) argue that dwelling depends on situational and local contexts. Thus, '"place" represents the particular to which are ascribed senses, practice, memory and desire, dwelling and movement' (ibid.: 8).

My fieldwork has revealed that the residents of Arnavutköy oppose not so much the construction of the bridge but, rather, the detachment from their 'place'. Many of them will have to move from their houses and those whose houses will not be torn down will have to adjust to a new reality that the construction of the bridge will impose. For example, Ortaköy, where the first bridge was built, faced the transformation of the seafront from a residential area to an area full of coffee-houses and restaurants. The area where the second bridge was built became undesirable for its residents, many of whom had to move due to noise and air pollution. However, none of my informants were willing to move from Arnavutköy or live in a different, changed Arnavutköy: 'I was born and raised here. I buy my bread here. My father had a shop in Arnavutköy. I breathe the air of Arnavutköy. It is a special place. Wherever I go, I always come back' (nineteen-year-old high school student); 'Arnavutköy is my life' (thirty-eight-year-old tailor); 'You can't stay out of this struggle. Your natural instinct pushes you to fight. It's like somebody is killing you ... This house is the same for generations, and it will not change. My grandson will grow up here' (forty-year-old businessman).

Of course these issues do not concern only the non-active ASG participants of Arnavutköylites, but for them it is the main issue. For the active and less active participants it is a significant, but not the most significant, reason which reinforces the opposition to the bridge. The residents of Arnavutköy do not want to leave the place they have inhabited most of all of their lives. The words of a resident describe this view:

> Arnavutköy is my home. Now when we finish I am going to go back to Arnavutköy and I feel very happy there. I walk very happily in the streets, I run early in the morning along the coast, I go to my barber and we chat a lot and he says, 'Why didn't you come? It's been a long time since I last saw you!' That sort of thing, it's my life. And like everybody else I would like to protect my life with the best possible means I have. That's why I fight. I am a school teacher by accident. I am *Arnavutköylü* (Arnavutköylite) first and then a school teacher. And it happens that the school I work in is not very far from Arnavutköy! But the important thing is that I would like to save the life and the world that I love (NAP, sixty-year-old academic).

ASG supporters

Finally, there are the supporters of ASG who are not residents of Arnavutköy but scientists and activists who provide ASG with scientific data, artists and public figures who help organise events and publicise the issues to a wider public simply because their well-known name or face is associated with the initiative. These people are what could be called the power brokers of the initiative. These issues are brought to the agenda by individuals who lend their expertise and legitimacy as scientists, activists or professional associations to ASG and mediate on the latter's account to public discourse.

In addition to environmental action groups (*Çekul Vakfi*, DHKD), ASG has been supported by scientists, members of the Chamber of Architects (*Mimarlar Odası*), artists, intellectuals and individuals from various parts of the country and outside its borders, since the issue (as already mentioned) has been picked up by international media such as the *Washington Post*, the *Los Angeles Times* and *Le Monde*. The co-ordinators of the master-plan project of Istanbul also support the initiative. The periodical publication of the Chamber of Architects (*Mimarlara Mektup*) very often mentions the issue and explicitly states its support to ASG. In a recent documentary prepared on their account, entitled 'City Crimes' (*Kent Suçları*), there have been references to the architectural 'crimes' that took place or are about to take place in Istanbul; the Third Bridge is also mentioned. A good number of artists and lay people have been supporting the movement against the construction of the Third Bridge; for example: Sezen Aksu, a very popular singer who actually shot one of her video clips (*Sarı Odalar*) in Arnavutköy as an act of solidarity to the Arnavutköy struggle; artists who volunteer to perform to the annual festival; and thirdly, the Alumni of the highly appreciated Robert College, the first private American High School of Istanbul whose buildings are in the area, launched an anti-bridge signature campaign.

The preservation and restoration of Arnavutköy's Ottoman architecture, the protection of its physical environment, the exposition of its historical and cultural heritage, the development of projects that constitute the area a leisurely walk district, the prohibition of cars in the heart of the neighbourhood and, finally, the protection of people's civil rights as well as the respect for people's opinion on their place of living everywhere – not only in Arnavutköy – are general ideals. They are not directly linked to the Third Bridge; they are rather legitimate issues to be discussed with official sites of power which do not take the focus off the Third Bridge even though they may not directly address it. In other cases, such as the documentary made by the Istanbul Chamber of Architects, the reference is explicitly made to legitimate the discourse in terms of architectural history and aesthetics.

Through activities such as the above, the supporters of ASG emerge as agents between dominant knowledge regarding environmental and cultural heritage protection and the rest of the public. This knowledge concerns environmentalism as expressed by international NGOs, the European Union which Turkey aspires to join and, as paradoxical as it may seem, by state authorities as well; and this is knowledge which the active participants of ASG are well aware, and, to a large extent, endorse. During the first months of the campaign – thanks to the help of its supporters – ASG, managed to include Arnavutköy in the ICOMOS 'heritage@risk' list and tried to find allies within the administration of the government. Lobbying efforts did not stop there. ASG participants informed national and international media, Turkish and international NGOs, and they have tried not only to convey their messages but also to incorporate the principles of these actors in their discourse. For example:

> If you build railways, if you focus on public transportation nobody would use cars. Using cars means more pollution ... I mean public transportation is cheaper, cleaner, much more comfortable, if there is good public transportation and I still don't understand why they don't focus on that. I mean, building a bridge is much more expensive, if you use the sea is less expensive. You build the *iskele* (ferry station) and all you have to do is buy a few boats and you put more regular, every half an hour let's say. It's going to be much cheaper and in the long run it is going to be much better (fifty-two-year-old architect).

> May the *UNESCO* project helps Arnavutköy to remain old. This way it will be a tourist place (fifty-year-old journalist).

> The ministry might have done research but we also have scientific research papers published from scientists and NGOs which prove that the atmospheric pollution caused by the cars that cross the bridges is enormous (fifty-six-year-old architect).

> What Arnavutköy people have done you may heard, we have reached ICOMOS [...] ICOMOS has listed Arnavutköy as 'heritage@risk'. So there is international support (fifty-two-year-old engineer).

As Stephen (2005) notes, social movements draw their arguments on social locations that are comprehensible to those they oppose to. Kasmir (2005) made a similar argument referring to a case of conflict between the workers of an automobile plant and the administration of it. She claims that the workers' outrage was framed in the very language of corporate hegemony. Likewise Kousis and Eder (2001), referring to European environmentalism, write that elite actors crosscut national publics and transgress national boundaries. Organised as 'elite publics' can be industrialists, NGOs, cultural entrepreneurs (intellectuals), media and professional groups. These groups explicitly use Europe as an opportunity structure to form a European association, a European network, or a European image. Following this, I claim that ASG uses dominant environmental discourse in order to create a legitimate environmental agenda, the principles of which have already been accepted by the wider public. A fifty-six-year-old architect said: 'ASG struggle is not violent and aggressive. They just say no. They struggle through the government means which is very important. So, it is not a protest like an outrage and rebellious which is very, very good. It is systematic, slow, and they are using the government tactics.' ASG principles through its supporters made their way out to the public and managed to be adopted by the public opinion both in and outside Arnavutköy. As in the case of the media, the supporters of ASG have made the issue public and have helped in putting the issue in the agenda of transnational bodies such UNESCO that have dominant status and give legitimacy to the anti-bridge campaign.

Conclusions

The ASG is clearly an environmentalist group; however, its distinct cultural and political characteristics, such as demands for democracy and for respect for human rights, take the initiative beyond a debate purely about the protection of nature. The present chapter has discussed both the environmentalist and the transenvironmentalist features of ASG. As an environmentalist group it is concerned with mainstream issues, such as sustainability and protection of cultural heritage. As a transenvironmental group the ASG promotes positive action not only for the benefit of the natural environment, but for betterment of society as well. In one of our discussions over tea, Ayşe told me:

> After the 1980 *coup d'état*, the situation was such that my husband and I went to live abroad because we were leftists. There, we were fortunate enough to make some money. When things were better we decided to come back to Turkey and with the money we had earned we bought this house. Now, if the bridge is built we have to leave again.

For Ayşe, leaving her house in Arnavutköy represented much more than an unpleasant move. It meant that, once again, the Turkish state could force people to leave their lives and, as another informant told me, referring to ways that politics are being practised by the Turkish state, 'That's enough!' (*yeter*!)

The story of the ASG is one of an urban conflict between the state and a neighbourhood initiative which brings together people from different backgrounds and unites their issue-specific interests and goals. The dissemination of their beliefs through their clever use of the media and internet technology, open forums in the community, public celebrations using culturally understood symbols, and celebrity spokespersons who manage to radiate the claims of ASG outward from the neighbourhood to the city and transnationally in Europe. Finally, the politicisation of the campaign through the contact with political parties, the state and the efforts to establish a local fire department testify that ASG is a collective action which makes urban demands in many levels.

Given the above, I define the ASG as an urban protest group – the particular character which I had the opportunity to experience through participant observation. This character concerns the identities of the individual participants, the people with roots in Arnavutköy, the supporters with sympathies rooted in the ideals of the ASG. As Barbara Bender (1995: 2) suggests, 'the way in which people – anywhere, everywhere – understand and engage with their worlds will depend upon the specific time and place and historical conditions. It will depend upon their gender, age, class, caste and on their social and economic situation.' In her analysis, she stresses that there are various interpretations of and connections with the world according to people's social identities, which are various and coexisting as in the case of ASG.

An additional objective of the chapter was to distinguish concepts such as secularism, localism and global citizenship and ideological identities such as leftism. The ASG is a group of people consisting of a core and a mulit-layered periphery with a core of the active participants and the periphery divided into the less active participants, the non-active participants and finally the supporters. This conclusion or model emerges from observations of the different subgroups participating in the initiative, which according to my research are formed by different identities which in turn constitute decisions for the opposition to the Third Bridge. All these identities have political references, and, although they constitute different subgroups within ASG, they share a common cause: to prevent the construction of the bridge; not just because it is going to be placed in their backyard. More important, if they succeed in having the decision annulled, they will have managed to render Arnavutköy everyone's backyard, the issue of public transportation of Istanbul everyone's issue and the problems of democracy in their country, everyone's problems. An excerpt from a paper written by two ASG participants relays the logic and emotion of the issue: 'ASG's final victory shall come when the central govenrment in Ankara will decide to permit the decision for a Third Bridge project over the Bosphorus to be taken by the citizens of the Municipality of Istanbul' (Danışman and Üstün 2003: 8).

All of the subgroups that participate in ASG share a common objective which is to prevent the bridge from being built. This objective fuses with wider beliefs regarding polity, lifestyles, science and life in general and along with the resources available to ASG, synthesise a legitimate agenda of discourses against the Third Bridge. The ASG participants claim that the Third Bridge issue is the result of years of bad development policies. Their demands for better public transportation, creation of a local fire department and for the revitalisation of ferry transportation illustrate that ASG is a lot more than a NIMBY protest. As Rootes (1999) underlines, involvement in NIMBY struggles can be proved profoundly educational. People who are involved in such campaigns develop a wider understanding of the environment and are able to criticise not only the problems in their backyard but also the problems in other backyards. Indeed ASG participants in their effort to strengthen their arguments started researching their place, understood why they did not want to leave it and that search uncovered a whole context within which they could contextualise their arguments. This context refers to scientific environmental awareness and the realisation that the decision for the construction of the bridge had political implications. These facts suggest that ASG is a social movement – or, better, a social movement organisation – to the extent that it identifies its claims with demands of wider social movements, which will be the focus of the next chapter.

Figure 1. One of Arnavutköy's narrow streets (*sokaklar*) with its famous timber houses (photograph by A. Voulvouli).

Figure 2. Arnavutköy after a snowfall (photograph by A. Voulvouli).

Figure 3. The Bosphorus Bridge viewed from Arnavutköy
(photograph by A. Voulvouli).

Figure 4. ASG participants during the Marmaray Inauguration holding banners and placards stating, 'Let's pass through the tunnel and cancel the bridge' and 'One tunnel equals ten bridges' (photograph by A. Voulvouli).

Figure 5. A wall of one of Arnavutköy's timber houses with the statement: 'The Third Bridge will destroy our Arnavutköy' (photograph by A. Voulvouli).

Figure 6. ASG participants during the Marmaray Inauguration Ceremony with banners. On the left, 'Yes to the Tunnel' (in Turkish) with 'yes' in English, French, Greek and Russian – 'Yes, oui, nai, da'. The banner on the right reads, 'The tunnel equals ten bridges'. Both banners are signed by *Arnavutköy Semt Girişimi* (Arnavutköy District Initiative) (photograph by A. Voulvouli).

Figure 7. Two banners hanging above Beyaz Gül Street in the heart of Arnavutköy: 'Third Bridge: Hayir, No Nein, Non'; 'I need the tube tunnel' (photograph by A. Voulvouli).

CHAPTER 4

The ASG in the Turkish Transenvironmental Campaigns Context

Introduction

Following the foundation of the Turkish Republic in 1923, the country experienced radical institutional, economic and cultural changes. The transition from a religious empire to a secular Western nation-state was accompanied by changes in the language, attire, gender relations which created drastic transformations in a relatively short period of time. After the death of the founder of the Turkish state, Mustafa Kemal Atatürk, in 1938 and the introduction of a multiparty political system in the 1950s, the Turkish political landscape was shaped by civil unrest, repetitive elections and *coups d'état*. The opening of the country both economically and culturally to the West, the marginalisation of Islamism, the influence of Kemalist ideology and policies of the Turkish army contributed to the formation of a peculiar kind of heavily centralised statist society in constant communication both with the capitalist West and – during some periods – the communist East; a statist economy with neo-liberal characteristics, especially after the 1980s.

In her book *Faces of the State*, Navaro-Yashin (2002: 12) claims that under the construct of 'Turkey', 'the capital city Ankara was founded, railroads were designed, schools were built, the postal system was changed, measurements were reformulated, the legal system was reframed, borders were patrolled, soldiers were conscripted, and women were employed in the labour force'. In her opinion the presence of the 'political' in Turkish 'public life' is too strong to be ignored by students of Turkish culture, even those who study non-state contexts. Following these views, I will attempt an analysis of Turkish non-governmental organisations (NGOs) by examining ethnographic material, as the consequence of the influence of hegemonic political and economic processes on the civic actions of Turkish people. Such groups are influenced by state-level institutions, and often function in opposition to what they proclaim to be; that is, structures separate and distinct from the state. On the contrary, they function in accordance with and some times even as an aid to state policies. Today in Turkey there are almost 3,000 NGOs focusing on a number of issues such as human rights, democratisation, peace, environment, Westernisation, Atatürkism, nationalism, and Islamisation (Keyman and Içduygu 2003). Some of these belong to a category of NGOs which oppose the state's choices while others belong to the category of NGOs which function as facilitators of state policies; but as either challengers or as facilitators of state policies, both use these policies as their main point of reference.

In this chapter I refer both to cases of environmental organisations which act as facilitators for the implementation of policies advocated by the government and to local environmental campaigns. The former are civil movements, in the sense that they are non-governmental which however, in various ways, 'rally around and organise for the state' (Navaro-Yashin 2002: 135). The latter are NGOs which are motivated by opposition to state policies focus on environmental, religious and human rights issues, identify their claims with wider social movements' claims and, like the ASG, are part of them. Economic and media globalisation have given rise to identity movements which have developed a hybrid character – for example, religious environmental, ethnic environmental and women's environmental (Ignatow 2005a). The case studies I present concern social mobilisations which draw on the same genre of arguments and, even though triggered by different events, their character is similar. They all demand policy makers to hear the argument; and they all feel that their claims and demands are legitimate. This discussion aims to introduce the Turkish scene of environmental campaigns to which the ASG – with its own distinct characteristics – belongs and by which it is influenced.

The Turkish version of environmentalism: Turkish environmental organisations

Even when environmental concerns are universal, each nation and each culture mobilises differently in order to protect its environment (Kousis and Eder 2001). 'The National peculiarities of environmental movements clearly testify to the persistent impact of national cultures and political structures and bear the imprint of national policies' (Rootes 1999: 304). Even though solutions to local issues demand transnational advocacy, transnational movement organisations are not ready to fulfill such an obligation. This explains why environmental activists' political actions – even when they have a global understanding of environmental problems – are defined by their national and cultural context (ibid.). In this sense, it would be useful to review the character of Turkish environmentalism both as far as state and non-state bodies are concerned.

The history of environmental organisations in Turkey can be traced to 1909, when the Law of Associations (*Cemiyetler Kanunu*) was issued. After that, new organisations made their appearance and by the 1930s, a number of environmental organisations emerged such as the Animal Conservation Association, the Island Reconstruction Association, the Embellishment of Çamlica Association, the Mountaineer Club, and the Turkish Association of Foresters. While these new organisations were coming to life, serious environmental problems also emerged such as the inadequacy of infrastructures in big cities, the rise of squatter housing due to rapid industrialisation, and migration to large cities.

After the establishment of the multiparty system, control over the associations and the political participation of the citizens was lessened thus causing the emergence of new social groups that have represented different interests in and views on the public sphere. The Turkish Association for the Conservation of Nature and Natural Resources was established in 1955 as a response to the environmental problems Turkey was facing and its foundations has been considered one of the first achievements of Turkish environmentalism. Its founders were forestry engineers, bureaucrats, and scientists who aimed to prevent forest destruction and preserving soil and water resources (Özdemir 2003).

In 1961 the term 'environment' (*çevre*) appeared for the first time in the Turkish Constitution. According to Article 49, 'everyone's physical and mental health should be protected', and, based on this argument, environmental institutions have been developed and environmental measures have been taken. Ever since, the Turkish government became member in many international environmental institutions and a plan based on the Third Five-Year Development Plan (1973–7) incorporated environmental protection regulations in the corpus of existing law. In 1978, the government formed the Undersecretariat of Environment and, in 1982, the Turkish Constitution included a number of articles regarding environmental protection which led to the 1983 Environment Law. Accordingly there was the proclamation of National Park Law, Law for Protection of Cultural and Natural Wealth, the Law to Protect Coastal Zones, Construction Law, the Municipalities' Law, Law to Protect General Health, Law to Encourage Tourism, Law of Forestry, Law of Water Products, and the Decree on Using Agricultural Lands. Finally, in 1991 the Turkish Ministry of Environment was founded (Arsel n.d.) and within this framework, the constitution of environmental organisations was encouraged by the state, which nevertheless held the responsibility for the protection of the natural and cultural environment (ibid.).

Today, major environmental problems of Turkey are the urban environment (air quality, water supply and waste water, as well as solid waste management) (Ignatow 2005a), the natural resources management (water resources, soils and land, forests, biodiversity), the marine and coastal resources, cultural and natural heritage and natural as well as man-made environmental hazards (Özdemir 2003). Another environmental problem which is also in the ASG agenda is noise pollution; as Ayşe (see Chapter 3) told me, in an effort to substantiate the effects that the construction of the bridge would have on Arnavutköy. 'Living under the second bridge, some people did not feel the earthquake [the 1999 Marmara earthquake] due to the fact that their houses are shaking all the time due to the traffic on the bridge!' Noise pollution is especially encountered in big cities, damaging human auditory health and perception – upsetting physiological and psychological balance – and lowering work performance. The main sources of noise pollution are traffic, places of entertainment, building and road construction and industry (ibid.).

Following Kousis and Eder's (2001) assumption that the type of mobilisation each society chooses to follow depends on the type of impact human action has upon nature, one might expect that environmental organisations in Turkey would have organised their activities around the issues mentioned above. However, according to scholars of Turkish environmentalism there are other parameters to be taken into consideration; for example, Özdemir (2003) underlines that the development of environmental awareness in modern Turkey has coincided with the development of democracy and human rights claims. In addition, Turkish environmentalism is influenced by Western environmentalism in terms of developing a local voice with international implications. As Ignatow puts it: 'The bright image of Western modernity guides Turks' efforts to protect their country's natural environment, too' (2008b: 55). An example of such a fusion is witnessed at the mass local demonstrations opposed to certain developments or aimed at the preservation of parks in the capital. Small groups like ASG are also influenced by this atmosphere. One of my informants told me:

> I was in Boston about two years ago and I had to travel by bus. The main road as you come out of Boston towards the south, to Cape Cod, is separated by two lanes. The private vehicles that were carrying only one passenger were obligated to follow lane number one whereas vehicles with more than three passengers could circulate easier. So, what happens is that in a neighbourhood three or four families come together and they share a car. Our bus could also go to the less jammed lane. So, even in places like America where cars constitute symbols of personal freedom you have a limited right to use them. If you are using them in a way that affects other people's quality of life then your freedom stops being a freedom. So, I think in this country people have to realise that there is a good public transportation system and that you are able to go to your job quickly and in much comfort by using public transportation. There should be ways of making you not to use your car. For example in London, if you take your car into central London, you pay a very high parking ticket ... Why do these people do that in Europe? Why don't we learn from their mistake? They have already made the mistake! Unfortunately the argument, 'This is my car, this is my freedom, I can do whatever I want with it,' is wrong. You cannot do anything you want with it if you are harming other people, if you are affecting their quality of life, and I think that a lot of people are now realising this (LAP, sixty-year-old architect).

Muslim environmentalists in Turkey draw on Islamic values in order to raise environmental consciousness. Furthermore, the environmental cause has brought people together from different religious and other social backgrounds such as leftist groups, secularists and religious organisations, who demonstrate for the protection of the environment. In this sense, Turkish environmentalism does not have a single identity. It is a realm where different identities co-exist and discover their interdependence, first with the natural world and secondly with each other.

Another issue to be considered is that responsible environmental behaviour is determined by knowledge, which means that education can play a great role in environmental awareness. Turks who are informed about the environment demonstrate a more responsible behaviour and join environmentalist organisations more often than uninformed individuals (Kasapoğlu and Ecevit 2002). Muammer (n.d.) makes a similar point when writing that environmental commitment is associated with education and occupation. Highly educated and highly prestigious professionals are more committed environmentally than lower-educated and less prestigious professionals. However, since there is no environmental education curriculum in the Turkish educational system, environmental awareness is the result of numerous environmental campaigns in the country (Tuncer et al. 2005).

Arsel (n.d.) mentions that such campaigns assume the role of the official government, since its presence in the formation of environmental policy is in many instances insufficient. Examples of such protests are the campaigns against the development of thermal power plants (Gökova, Aliağa and Yatağan), gold-mining (Bergama), nuclear power plants (Akkuyu) and hydroelectric dams (İlisu, Tunceli).

Environmental associations and transenvironmental campaigns

McCormick (1995: xi) argues that environmentalism generated new bodies of law, hatched new political parties, encouraged rethinking of economic and social priorities, and became a central issue in international relations. However, the environmental movement has been characterised by a lack of homogeneity. The green literature is, for example, divided between the technocratic environmentalists who see environmental problems resolvable by the 'technical fix' within existing economic and social practices, and the ecologists who see such problems as embedded in economic and social structures and practices and resolvable only by changes at that level (Martell 1995: 7). Yearly (1993) claims that difference is expressed also by NGOs; on the one hand, there are the conservation groups with a pure scientific orientation and, on the other hand, there are the environmental groups which have their roots in the 1960s and 1970s and which relate environmental degradation to the social structures of Western society.

Similarly, Turkish environmentalism consists of three groups of organisations. The first is concerned with a broad but ineffective official sector. The second is the civil environmental movement (foundations, associations and cooperatives formed by the private sector) associated with the first group and, in some cases, shows a technocratic tendency when dealing with environmental groups. Finally, there is a third group which is non-civil in terms of a political agenda (the Green Party, for example) but is relatively independent from the first group, yet also ineffective. Within this framework there are four distinct groups: the Greens, the radical environmentalists, the conservationists and the Islamist environmental groups (Özdemir 2003).

The Turkish Green Party was founded in 1988 under the influence of the Western Green movements by members of marginal groups, such as feminists, atheists, homosexuals, lesbians, anti-militarists and others. All these marginal groups proclaim their autonomy from the state and accuse other environmental groups of this collaboration. According to the Turkish Green Party, environmental degradation is a result of the Enlightenment and this is why it rejects the concepts stemming from Enlightenment and modernity. Some of these concepts, such as the theory of linear development, Western cultural and educational norms have been accepted by many modernist Turks, whereas the Greens hold a worldview which is more holistic and humane. Because of the profile of the founders of the Green Party many people were sceptical about the real aims of it. They thought that the Green Party was using environmentalism in order to legitimise its other characteristics. In this landscape even the Turkish Foundation for Environmental Protection announced their scepticism regarding the

Green Party and its founders. Even though the party was banned six years after its establishment, it had already contributed to the development of environmental agendas in other mainstream parties.

The second group of environmentalists is composed of those who claim that democratisation and environmental protection go hand in hand. According to them, environmental problems are the product of undemocratic governance; hence, environmental problems are democratic problems that can be solved through political empowerment and participation. An example of these types of associations is the SOS Mediterranean Association group which was founded in 1999. At the beginning of its establishment, it was a division within the Green Party that aimed at the protection of Mediterranean ecosystems, the termination of the ecological crisis of the region and the advocacy for peace in the Mediterranean. Tourism and industrialism, along with polluted rivers, endangered species and destroyed historical sites are, according to them, among the most serious problems of the region. According to this group of environmentalists, harmony between man and nature can be accomplished through Green philosophy and ecological policies. The aims of the association as mentioned in its website concern: 'The information and awareness raising campaigns on specific ecological/environmental key issues for the protection of natural resources, the organisation of meetings and conferences centring on specific and actual ecological issues, the publication of magazines, books, leaflets promoting environmental education and public awareness and the Cooperation with National and International, Mediterranean and European organisations' (http://www.evonymos.org/english/medit_sos.html).

Another group of organisations are the conservationist associations which are characterised by their lack of political and radical discourse. Two of the most prominent organisations representing this approach to environmental protection are the Environment Foundation of Turkey – EFT (*Türkiye Çevre Vakfi*) and the Turkish Foundation for Combating Soil Erosion, for Reforestation and the Protection of Natural Habitats (TEMA). The first group was founded in 1978 as a non-governmental, non-profit, voluntary and independent organisation carrying out its work in accordance with the principles of the Civil Code of Turkey. It is an organisation with liaisons with the UN, the EU, the World Bank and the

Organisation for Security and Cooperation in Europe. Its strategy is to carry out research and raise public awareness. The environmental literature in Turkey has been increased because many books on environmental issues have been published by EFT.

The second organisation in the conservative group, TEMA, was established by a Turkish businessman, Hayrettin Karaca, in 1992 and focuses on the prevention of soil erosion, the protection of the natural heritage, the preservation of biological diversity and the organisation of environmental courses for the public. According to its founder, environmental problems are associated with political and economic systems, consumption patterns and modern civilisation. Therefore Turkey's environmental degradation causes social problems such as migration, underdevelopment and starvation. Although TEMA was established by businessmen, it is critical of modernist development and consumption patterns. TEMA focuses mainly on raising public awareness about desertification resulting from widespread soil erosion in Turkey. This goal involves informing and lobbying governments, national representing parties, educational institutions and private organisations (www.tema.org.tr). It also carries out pilot projects on rural development, rangeland rehabilitation and reforestation. Furthermore, it collaborates with state institutions such as the military and several ministries in order to implement a variety of environmental projects. Like EFT, TEMA publishes books on environmental issues.

Finally, the involvement of Islamist groups in modern environmental debate is a recent phenomenon of the last two decades. Muslim thinkers and scholars have written much about the environment. They see environmental degradation as a result of modernity in general and modern, secular culture in particular (Ignatow 2005a). According to them, the Islamic world view and everyday practice could contribute to the protection of the environment. 'Islamist environmentalists tend to be Islamists first and environmentalists second' (Ignatow 2007: 72). However, the Turkish political system, which does not facilitate citizens' involvement in the affairs of the state in addition to the secular structure of the state that does not tolerate any organised civil activity for religious groups, has acted as hindrance for the establishment of Islamist environmental groups. This is why Turkish Muslim environmentalists' activities have been restricted to the publication of scholarly and popular works on environmental awareness (Gürsel 1989; Toprak 1993; Bulaç 1995; Özdemir 2003) and to the organisation of conferences, panels, workshops, and symposia.

On the contrary, political Islam has extended its activities to policy making, especially after winning the local elections in 1995. The Islamist-oriented mayors of big cities, such as Istanbul and Ankara, gave priority to environmental issues and they are considered as having been very successful. However, they have been accused by their political enemies that they have been reactionaries who have been using the green colour not as the symbol of environmentalism but as the symbol of Muslim civilisation. In this sense, according to secularist circles, the Islamist politicians have been using environment as an excuse for the promotion of their religious values. Nevertheless, as even Islamist environmentalists admit, Islamist environmentalists have not succeeded in being influential on Islamist or national politics for various reasons (Ignatow 2008a). In contrast, Alevi environmentalism that focuses on specific issues seems to be more influential.

The religious minority of Alevis in Turkey constitutes 20 per cent of the Turkish population (Shankland 2003). During the Ottoman reign, Alevis were overshadowed by the dominant Sunnis and this is why they have warmly endorsed the Young Turk revolution. However, Alevi groups were also suppressed by Turkish leaders (Ignatow 2008b) and even Atatürk himself turned against them. The paradox is that Alevis never hold Atatürk responsible for their oppression. On the contrary they include Atatürk among their principal saints, along with Ali, the son-in-law of Muhammad, and Sufi mystic Haji Bektaşi (Shankland 2003; Ignatow 2008a). Due to the residence of Alevi in predominantly Kurdish regions, they have been considered a Kurdish subgroup with leftist beliefs (Shankland 2003) and a liberal Islamist minority with a strong love of nature (Ignatow 2008a). Shankland (2003) claims that Kurdish as well as Alevi social organisation tends to oppose central authority. Therefore, as other non-governmental groups during the 1980s and 1990s, Alevi Muslims have come forward asking for recognition from the Turkish state and society and a number of Alevi organisations have been founded (ibid.). Ignatow (2008a: 10) claims that 'Alevi environmentalism combines traditional Alevi beliefs with ideas drawn from science and global civil society'. For Alevis,

Alevi-Bektaşi philosophy and theology is compatible to sustainable form of living as it respects the balance of nature. Alevi leaders, such as Celal Arslan and Ayhan Aydin proclaim Alevism's relation to environmentalism and environmental activism, conspicuous in the environmentally themed architectural competition for Cemeviler (Alevi houses of worship), organised by the Cem Foundation in 1996, and tree planting campaigns all over the country (ibid.).

As Ignatow (2007: 64) claims, 'Most of Turkey's large environmental groups, like EFT and TEMA have strong links to the state.' The above environmental groups have allied with liberal industrialists, the EU and the United nations (Ignatow 2008a). EFT for example, is linked with international organisations such as the United Nations, the European Union, the Organisation for Security and Cooperation in Europe and the World Bank. At the same time, the UNEP National Committee for Turkey and the EURASIA Environmental NGOs Information Center are incorporated within the legal framework of EFT. The Foundation is also in cooperation with the Centre for Our Common Future, located in Geneva (www.cevre.org).

The Green Party and the Islamist mayors have been involved directly in politics while TEMA, EFT and the SOS Mediterranean Association Group are officially acknowledged environmental organisations, even though of different orientation, and they have alliances with state institutions. Several examples can prove such alliances. For example, in 2007 TEMA organised a workshop to decide on the process of producing National Red Data Books in Turkey. This workshop was coordinated by the TEMA Foundation in partnership with the Turkish Ministry of Environment and Forestry. As the website of the foundation informs us, in the workshop

> experts from the Turkish Ministry of Environment and Forestry, academics, NGO representatives, and IUCN representatives gathered to present the current status of the national red data books, to discuss the process for finalising, maintaining and updating red data books of Turkey. Participants discussed how national red data books can assist in implementing national policies on nature conservation. The workshop concluded to establish working groups on taxonomic classes, and a steering committee to lead the process. Experts discussed how to manage and update data on red list species and how existing biodiversity databases can be incorporated

> into this process. The Ministry of Environment and Forestry was present in all the discussions. The workshop has significantly contributed for creating a consensus on the process and initiated the preparation of many national red lists in close future (www.tema.org.tr).

In addition, as mentioned above such organisations play a complementary role to that of the state. For example, as EFT informs us:

> In 1978, when the concept of environment was not too well known, EFT pioneered for it to be taken into consideration by the government. As a result of its efforts from 1980 through 1982, Article 56, which deals with the environment was included in the Constitution. On the other hand, EFT prepared a draft Environment Law, insistently followed it through for two years and played the greatest role in its acceptance. Furthermore, also in 1981, EFT prepared the first environmental profile of Turkey, being the first volunteer organisation in the world to publish such a profile. Updated editions of the profile are published every two years by EFT who has undertaken to render this service which is carried out only by governments in many other countries (www.cevre.org).

TEMA has also aid in implementing projects such as reforestation, afforestation, carbon sequestration and biodiversity protection projects in areas subject to erosion, in order to regain the green cover, wild life and ecological balance. The foundation, with the support of the Ministry of Environment and Forestry, has planted 2.2 million saplings over an area of 2,350 hectares under twenty-five different reforestation projects through the sponsorship of individuals and companies (www.tema.org.tr).

Both orthodox Islamists and Alevis have links with international organisations such as Greenpeace and Friends of the Earth. In addition, Alevi Muslims are being supported by Alevis of the diaspora who live mainly in Europe and Australia (Ignatow 2007). In this way, these groups have found ways to institutionalise their environmental beliefs.

On the other side of this model we find local environmental struggles that oppose to environmentally harmful developments. Even though Alevi associations have been part of such campaigns as we will see below, these groups constitute local environmental protests which are not official environmental organisations. Arsel (2003) claims that the first movements protesting a development in Turkey were NIMBY-style protests until the

Bergama resistance, which has been the turning point of environmental campaigns. The Bergama protest altered the character of these types of mobilisations through the transmission of information of their struggle through the national media.[1]

The Bergama protest

A much studied and discussed protest is that of Bergama. It concerns a conflict between the residents of the Bergama area near Izmir and the Eurogold Corporation over the establishment of a gold mine in the area which is using cyanide to extract gold. The majority of the residents of the area have strongly opposed the existence of the mine due to the potential harmful side-effects that cyanide has on public health. The controversy created by this case led to the Turkish High Court (Öncü and Koçan; 2002; Arsel 2003; Özdemir 2003; Beşpınar-Ekici and Gökalp 2006). More specifically, in 1989 an Australian based multinational company known as Normandy Mining was granted a permit to start the Eurogold Project. The company made a strong public campaign aiming at presenting the project as a positive contribution to the area's socio-economic development. To demonstrate its considerate intentions the consortium announced its plan to construct a mosque and a wedding hall in Ovacık village where the mine would be based. Touched by the goodwill of Normandy some villagers agreed to sell their land to the corporation at a considerably higher price than the market rate at the time. Another group of villagers saw the mine as a positive development for the area as they perceived it as new source of employment (Öncü and Koçan 2001).

As required by Turkish legislation, in 1991 the company submitted an Environmental Impact Assessment Report (EIAR) of the gold mine to the Ministry of Environment. The EIAR asserted that cyanide usage is

1 Furthermore, the Bergama activists have sought allies in the international environmental community. Ever since, environmental protests follow similar patterns of internationalisation.

legitimate and not harmful to the local population. However, Normandy's history on environmental issues was infamous to academics, scientists, researchers and politicians who publicly claimed that cyanide leaching was, in fact, dangerous. This information along with the warnings received by international environmental organisations and an incident during the preliminary drilling for the construction of the mine that made the water of the area undrinkable for four months (Sachs 1997) initiated public reactions against the mine which, in turn, reinforced inquiries by the municipality regarding the cyanide leaching method (ibid.). An Environmental Executive Committee was formed with informal organisation. This committee made the issue known both nationally and internationally. For the locals it was not simply an environmental problem but a citizenship issue in terms of people's right to resist and protect their welfare and livelihoods. The extensive participation of women, the non-violent approach, the creativity, persistence, and determination of its protesters, and the occasional use of shocking and non-traditional forms of protest, including a nude protest, made the Bergama struggle widely known and appreciated (Özdemir 2003).[2] The Bergama citizens reached the Constitutional Court of the Republic of Turkey appealing to their constitutional rights, in terms of living in a healthy and balanced environment. The Court revoked the permit given to Normandy on the grounds that the mine breached the right to a healthy environment and would pose public health to a severe risk (Sachs 1997; Öncü and Koçan 2002).

The Bergama protest has been interpreted in various ways. Firstly, it has been examined as a case of multilayered governance; that is, as a locus in which two main political levels meet – the national and the transnational level (Öncü and Koçan 2001). Due to the government's hesitation to adopt the decision of the Constitutional Court, the Bergama villagers related their demands to global agendas such as the efficacy of individual, citizenship and human rights in the world context, the structure of the international order

2 In April 1997, the military tanks were sent to the area to restrain the protest but the villagers made a peaceful demonstration in which 10,000 people and 1,000 tractors took part (Sachs 1997).

and major global social problems such as environmental degradation, disproportionate wealth and the power of transnational corporations. During their effort they came to realise that politics dominates justice, and that the economy dominates politics, thus they turned themselves into political activists with a global character (Öncü and Koçan 2002). Eventually, they were associated with NGOs, social movements and international institutions (European Court of Human Rights, the UN War Crimes Tribunal in The Hague). This entails a global agenda which questions the exclusivity of governments over political power. In this way, new actors become involved and define two types of democratic citizenship movements. The first type is about the mobilisation for the effective use of existing rights, and the second type concerns the mobilisation for the recognition of rights, which – although they are recognised on a transnational level – are not recognised on a national level. The Bergama movement has claimed both existing rights, at the national level, and the recognition of transnationally accepted rights (Öncü and Koçan 2001).

Secondly, the Bergama protest has been seen as a case of demands for democratic rights. The protest – combined with the Bergama villagers' democratic rights – transformed the latter into a problem on the public agenda, forcing the government to change its policies. The Bergama protest appealed to democratic citizenship through mobilising people to exercise their rights. Thus, the villagers acted as citizens aware of their rights and the parameters of the violation of the law, and in this way, they contributed to democracy. They also reconstructed their social relations by rendering their village into a public forum that mediated between the state and society. In this sense, as Habermas argues regarding the emergence of a 'public sphere' in opposition to the formal political system, the Bergama villagers constituted a 'public realm of reasoning private persons' that was partially secured by the existence and the effective use of constitutional rights and freedoms (Öncü and Koçan 2002: 12).

Finally, it was considered as a case of environmental activism. The peasants who were demonstrating against the mine and for the protection of the environment, a decade previously were demonstrating for new industries in the name of employment and economic growth (Özdemir 2003). In the era of globalisation, where the units of analysis must be regional rather than national, the Bergama case constitutes a critique on the methods of development found in many rural regions across the world. It is a case of environmental activism often encountered in Turkey, a fact that demonstrates its growing importance. However, environmental activism goes beyond environmental improvement as such. Thus, the Bergama protest can also be interpreted as a civil society development (Arsel 2003) and as a human-rights protest (Sachs 1997). As part of the environmental movement it is seen as 'a celebrating example of constitutional democracy and citizenship', and moreover, as 'a precursor of social projects critical of the instrumental rationality in the Turkish context'. The Bergama movement provided an opportunity to implement the right of social opposition and civil disobedience (Özdemir 2003).

Metropolitan Municipality of Istanbul versus Chamber of Architects

Another well-known campaign was the one against the so-called Tarlabaşı demolitions, a decision of the Metropolitan Municipality of Istanbul and its mayor at that time, Bedrettin Dalan. The campaign was mainly organised by the Chamber of Architects and opposed the plan for the transformation of one of the main streets of Beyoğlu[3] district of the city, İstiklal Caddesi (Independence Street), into a pedestrian mall and the opening of a parallel artery which would ease traffic congestion. The realisation of the project demanded the demolition of a large number of nineteenth-century buildings (Bartu 1999). As Bartu (1999) claimed, the dispute over the demolitions did not only concern the clash between conservationists and modernisers. It entailed a series of issues which at the time of the conflict became publicly critical political issues. Considering the time period over which it took place, a few years after the military *coup d'état* of the 1980, it is fair to say the Tarlabaşı demolitions were a means to express discontent for the general political situation in the country, making Tarlabaşı into a political symbol.

3 Also known as 'Pera'.

The İlisu Dam case

The İlisu Dam case centres on a decision taken by the government to build a dam in the area of İlisu which is situated in the predominantly Kurdish south-east area of Turkey. The objections had to do with the failure of the Environmental Impacts Assessment (EIA) of the project to prove that the area would not be affected by the dam. The locals also objected to the fact that the dam would flood the city of İlisu which is of historical and archaeological significance and that a lot of people would be unwillingly relocated. These factors constituted a threat to the local cultural heritage (KHRP 1999). Again, this case involves more than a clash between development and conservationism. The area in which the Dam was going to be built is mainly populated by people of Kurdish origin. This fact was incorporated in the campaign and the struggle against the dam also came to express the struggle of a subordinate group (Haynes 1999; Kousis and Eder 2001) – the citizens of Kurdish origin against the Turkish state. The case became noticeably public both inside and outside the country because of the involvement of international NGOs (e.g. Friends of the Earth) and local NGOs, such as the Kurdish Human Rights Project (KHRP), which is based in London but with representatives in the area.

The Tunceli campaign

A similar case is the Tunceli anti-dam campaign, characterised as an Alevi-Kurdish or Kurdish-Alevi campaign (Ignatow 2008a: 11). It concerns the opposition to a government project to construct eight dams at the Munzur River in the mainly Alevi- and Kurdish-populated Tunceli region of Anatolia, which is known for its socialist and anti-government groups. The third of eight planned dams has recently been completed but has not yet been filled, due to anti-dam activists who claim that if it does the water will flood the city of Tunceli and the roads, and will cut off already isolated villages populated by Kurdish immigrants during the Kurdish war in the 1990s (Ignatow 2007). Similar to the case of the ASG, the protestors of the Tunceli area have expressed their protest in an annual festival held in

Tunceli during August 2004. Even though the festival was not organised exclusively by the anti-dam protestors, most of the activities and panels were dedicated to the Munzur struggle. The Alevi involvement in this struggle was made evident through a theatrical show that depicted Alevi traditions concerning the sacredness of the river and the valley (Ignatow 2008b).

Conclusions: ASG in the Turkish transenvironmental campaigns context

María Pilar-García (1992), in discussing the mobilisation of environmental actors in Venezuela, concludes that the Venezuelan environmental movement is constituted by a network of organisations, each of which function in a different way. In this chapter, I have tried to illustrate that even though there are similar concerns between local environmental campaigns and organisations such as SOS Mediterranean Association Group and even TEMA and EFT, there are also clear differences. Firstly, while the first group of organisations consists of official associations with constituencies and members, local campaigns have unofficial status and their participants are not granted any membership. Secondly, protests like Bergama focus on a particular issue which they connect with wider socio-political demands such as democracy, human rights and sustainability. Environmental Associations such as TEMA and EFT have wider agendas which mainly focus on sustainable development and other 'official' prerogatives. An EFT statement illustrates this claim: 'EFT has introduced various new environmental concepts like sustainable development and biological diversity' (www.cevre.org). Finally, local protests are not hierarchical structures but *ad hoc* groups that enjoy the participation of any one who wishes to join at any moment of their struggle. In addition, they are not linked with political parties even though, as in the case of the ASG, opposition political parties have stated their supported in their struggle.

In brief, in this chapter I have tried to illustrate the differences between local environmental campaigns and environmental associations. Moreover – as will be discussed in detail in the next chapter – the division between environmental associations and transenvironmental campaigns challenges the notion of civil society and supports the consideration that Turkish environmental organisations do not belong strictly to the realm of civil society. As far as the Turkish context is concerned, the concept of civil society proves to be problematic as Turkish civil actors function either in co-operation with or against the state. Therefore, except from their differences, both these two sets of organisations have as their main point of reference the state and act either along with or against it. Even if the former protest against it and the latter work in parallel with it, they are both in interaction with it. Thus, instead of placing non-governmental environmental actors in a different realm from that of the state, a discourse about the points in their articulation should be addressed.

As mentioned earlier, the ASG is a response to and challenges political decisions. In this sense, it is part of the domain constituted by the political and the people; part of what Navaro-Yashin calls 'public life', where people and the state are examined not as different but as within the same domain. The groups similar to ASG, discussed in this chapter constitute some of the many examples of such campaigns and by discussing them the present chapter sets the context of which ASG forms part. Protests like those discussed here create a network of campaigns assuming the character of a social movement and ASG, along with all these other campaigns, is what Schönwälder (2002: 11) terms part of the larger universe of social movements that criticise society and propose different ways of practising politics.

In such protests citizens demand constitutional as well as international human and environmental rights. In an academic paper presented by two ASG participants to a conference on contentious politics and social movements held at the University of Crete in Greece, it was argued that 'the fight [over the bridge] turned from objecting a destructive bridge project into recognition of civic rights of citizens' (Danışman and Üstün 2003: 6).

> The agenda is not any more just the third bridge project, but they have widened their horizons and begun to discuss possible measures against environmental disasters likely to be caused from highly risky Russian oil tankers passing through the Bosphorus, or how to regenerate the tranditional fish species through projects such as sponsoring surface cleaning vessels for pollution collection, or organising civilian disaster management teams and first aid firestation in the neigbourhood to prepare for the eventual big earthquake (ibid.: 8).

In this sense, the ASG is about claiming new civilian rights in terms of actively protesting against an issue and not accepting it passively, as it used to be in the past, in the name of the republic (Beşpınar-Ekici and Gökalp 2006). To continue with the issues raised in the above paper, 'Arnavutköy residents are slowly becoming more active in shaping local and national political agendas' (ibid.: 8) and as Ayşe told me, 'All over Turkey, there is one ASG and one *Bergama* that have made the difference in politics.' Another informant said:

> All those damn' cases as well as the *Bergama* case, show that in Turkey things are no longer the same. People do not accept what others decide for them any more (LAP, forty-year-old psychologist).

Therefore it seems that the ASG, along with the other campaigns has created a loose network inside Turkey. At first glance this network could be characterised as an environmental network of organisations, a second glance suggests that it goes beyond a strictly environmental cause. Even though this book is not an ethnography of Turkish environmental campaigns and the material on which I drew my assumptions about these cases comes from studies and writings of other researchers, I could identify many similarities between their examples and the ASG one of the reasons for referring to their cases. In particular, the ASG, as emphasised by its participants, is one of many such campaigns which have been formed as an answer and response to environmental (but not only) policy-making. Collectively, these protests, along with the ASG campaign, suggest that environment-related conflicts between states and grass-roots movements constitute a form of social and political criticism and this in turn, suggests that their character depends on the surrounding political culture (Jamison et al.

1990: 2). Hence, more than environmental groups, these local protests are 'transenvironmetal' mobilisations. These campaigns, except from adopting dominant environmental prerogatives link their environmental concerns with wider societal issues such as democracy and human rights; thus transcending an otherwise strict environmental agenda.

As an analysis of the other cases supports, the ASG is part of a network of groups which compose the Turkish transenvironmental movement with all the variations that social movements – environmental or not – include. ASG, like the Bergama and the İlisu Dam campaigns, declares its interest in the environment in tandem with other issues, such as democracy, the effects of global economy on the local level, human rights and critique of political ethics.

This is the subject of the work at hand. The effort of a group of people, along with their supporters who range from individuals to institutions, to claim not only the right for a clean environment but also the right to decide for their place in the process. This is why their group belongs to the community of the Turkish transenvironmental campaigns and not to the environmental associations such as TEMA, EFT and SOS Mediterranean. ASG is a piece of Turkish social movements whose character is, in a large part, defined by the existence and the similarities between various unique, yet similar social mobilisations.

CHAPTER 5

ASG: A Social Movement Organisation between 'Civil Society' and the State

Introduction

This chapter presents a theoretical analysis of the ASG based on the ethnographic data introduced in the previous chapters. Consequently, it explores four areas: Firstly, as mentioned in Chapter 3, ASG has explicit environmental references, which means that the dispute over the Third Bridge should be examined as a case of an environmental conflict. Following the standpoint – as suggested by Milton – that environmentalism is a particular way of understanding the world and thus a type of cultural perspective (Milton 1996) it seems necessary to study different perceptions of the environment as different cultural perspectives and environmental conflicts as cultural conflicts. For Milton, 'environmentalism is part of the ways in which people understand the world and their place within it. It belongs to the sphere that includes people's feelings, thoughts, interpretations, knowledge, ideology, values and so on' (ibid.: 33). Moreover, as Jamison et al. (1990) claim, environmentalism is shaped by the political culture of specific places, and as Vassos Argyrou (2005: viii) reminds us:

> to take sides in the debate over facts concerning the environment would be to reduce environmentalism to a question of scientific objectivity. Indeed, to a large extent, it is on this assumption that the debate over facts is based – the assumption, that is, of rational actors making informed and rational decisions about the state of the world. Yet things are far murkier and convoluted. Environmental conflicts mainly emerge from the control over power. Saving nature is not about saving our modern-

ist culture but saving the power to 'define the meaning of the world for *everyone yet again*' (ibid.: ix).[1]

Following these assumptions, this chapter claims that environmental conflicts such as the one on which this book focuses are conflicts not over nature as such but over the power to define nature or, more correctly, to define what is good and what is bad for nature. All these reinforce the examination of environmentally oriented collective action like ASG as transenvironmental rather than environmental action (Kousis and Eder 2001); that is, an action in which the environmental cause articulates cultural, political and societal issues.

Secondly, based on this assumption (that *Arnavutköy District Initiative* is a collective action) or a 'social movement', as the press representative of the initiative told me in one of our discussions, in the pages that follow I explore the literature on the typologies of social movements in general and environmental movements in particular. According to Geralch (1999), a movement consists of a network of loose relationships between different groups. In Turkey – as mentioned in Chapter 4 – there are many such groups which, combined could be considered to constitute social movements. Therefore, in this chapter I focus on ethnographic accounts of collective action and anthropological analyses of such movements. These accounts often claim that the formation of extranational communities, such as the EU as well as globalisation, in terms of increased extensity, intensity and velocity of social interactions (Arsel n.d.: 7) transform such actions from national and local to transnational and global in terms of alliances and the use of certain discourses as the case of ASG suggests.

1 Therefore, scientific correctness or falseness should not be the focus of studies of environmental conflicts. Nature is not as 'natural' as it appears to be; to the contrary, it is a human construct and as 'cultural' as is culture. In this sense, 'nature' and 'culture' are not oppositions and should be treated as the same. That they are seen as two completely distinct concepts is made possible by a process of constant purification to which modern society subscribes (Latour 1993). Uncovering the hybrid character of nature and culture as nature/culture means revealing the impossibility of modern life – or to use Latour's precise words – 'Modern Constitution' – *to lend itself to the dualistic paradigm* of which paradoxically it is a product (Latour in Harbers 1995).

Thirdly, since analyses of social movements often situate them in the realm of the, so-called, civil society, the present chapter explores this concept. Taking the lead from theorists who reject the binary opposition between state and civil society, I claim that it is the study of their fusion rather than their separation that proves more fruitful in the analysis of social movements. Within this framework, anthropologists and other theorists of social movements focus on the ways local people seek to establish new relationships with the states in which they are included (Appadurai 1996). As the ASG case suggests, such movements can be seen as social conflicts over power that take organisational forms and express deeper rooted changes in society (Edelman 2001) and a variety of issues that illustrate how ASG and organisations like it combine a variety of characteristics and are not defined in terms of just one of these types of movements.

Environmentalism: Science, politics, movements

Jamison et al. (1990) speak about the four phases of environmentalism in Sweden, Denmark and the Netherlands. The first phase covers the period 1962–8 during which a new environmental consciousness was shaped through scientific criticism that developed into a public debate. The second phase extends from 1969 to 1973; during this period environmental groups were being initiated mainly by scientists. Those groups were different from the existing conservationist societies and were more activist in orientation. The third phase, from 1974 to 1980, has been referred to as the 'highpoint of environmentalism as a social movement'. During that period struggles against nuclear energy occurred and it has been recognised that environmental problems were primarily political and social. It was during this phase when ideological issues about environmentalism made their appearance in the agenda of environmental movements. Finally, phase four began in 1980 and continues to the present. This phase is characterised by the expansion of environmentalism and its coalescence into a structured social movement with organisations, leaders, factors that have brought internal conflicts.

The phases of Scandinavian environmentalism described above are fairly similar to the phases of environmentalism throughout the Western world. Therefore, below I will try to characterise Western environmentalism by three main time periods. The first period covers the nineteenth to the mid-twentieth century when environmentalism was a result of scientific understanding. The second and third periods extend from the 1960s until the 1990s, and from the 1990s to the present. During these periods the environmental movement assumed a political character and the environmental discourse assumed global dimensions in terms of the realisation that environmental problems demand solutions on a global scale.

Environmentalism as science

Environmentalism's origins can be found in the nineteenth century. In Australia and Africa the environmental movement was generated from reactions to the exploitation of natural resources (McCormick 1995). In Europe and North America, environmentalism was born out of concerns about preserving natural resources reinforced by improved scientific understanding and as a response to the threats posed to nature by increasing industrialism (Cladwell 1990; McCormick 1995; Raptis 2000). The first association for the protection of the environment was founded in 1854 in France, *The Zoological Association of Naturalisation* (Simmonet 1985; Flogaiti 1998). During that period, the first debates within the environmental movement appeared, between the preservationists who claimed that wilderness should not be exploited by any means and the conservationists who claimed that rational and sustainable exploitation of the natural resources could be achieved (Milton 1993 and 1996; McCormick 1995; Flogaiti 1998). Within this context the transcendentalism movement was initiated by the American literary writers and critics such as Emerson, Thoreau and Muir. Their arguments connected the deterioration of natural environment to the deterioration of the quality of human life (Flogaiti 1998).

Among the achievements of the environmental movement was the enactment of measures for the protection of entire areas of land (Raptis 2000). The first National Park, the Yellowstone Park, was established in

Wyoming, United States of America, in 1872 (McCormick 1995; Emberlin 1996). This date is considered very important for the history of the preservation of the environment because it was the first time that a whole area was protected by law (Flogaiti 1998). The example set by the United States of America was followed by other countries, mainly after the Second World War, and is still being observed (Sakellari 2001). For example, in Turkey thirty-five National Parks have been founded since 1958, along with eleven nature preserves and four special areas of environmental protection. Finally, eighteen sites are candidates for being registered in the World Heritage Convention which, along with the nine sites already registered bring the total to twenty-seven throughout Turkey (www.allaboutturkey.com).

Environmentalism as politics

For Guha and Martinez-Alier (1997), early environmentalism was the result of the industrial revolution. Later on, both in Southern and Northern nations environmentalism used the language of class and race as a base for organising. For example during that period social conflicts with an ecological content emerged, such as: struggles against environmental racism (dumps in minority communities), toxic imperialism (waste disposal in poorer countries), ecologically unequal exchange (based on prices which do not reflect local externalities), the North dumping subsidised agricultural surpluses in the South (to the detriment of small farmers there), and biopiracy (corporate appropriation of genetic resources without recognition of peasant or indigenous intellectual property rights).

Modern environmentalism has its roots in the alternative political culture of the 1960s which was influenced by ideas of participatory democracy and non-hierarchical organisational structure (Jamison et al. 1990). During that period, 'the environmental movement went beyond the plight of nature challenging the basic assumptions about human attitudes to the entire human environment' (McCormick 1995: 53). Post-materialist approaches examined nineteenth- and early twentieth-century conservation campaigns as projects of enlightened elites or even of a gentry overwhelmed by industrialisation. Northern environmentalist movements claimed that appreciation of natural amenities was the result of urbanisation and affluence. Finally, Grove (in Edelman 2001) explains the rise of environmentalism in Europe as an outcome of the encounter of the Europeans with the tropics and of their devastation caused by rapacious plantation economies.

The 1970s saw the rise of green political parties and many already existing, centre-left political parties in the West began to introduce environmental issues in their agendas and platforms (Marangudakis 1999). The 1970s are also remembered by environmentalists for the 1972 United Nations Conference on the Human Environment held in Stockholm (Einarsson 1993). In the 1980s environmental policy was influenced by the creation of environmental NGOs and during this period it became conspicuous that the rich and the poor nations had very different understandings of the environment and its place in development. During that decade a shift in the way that environmentalism was expressed was observed, since, by then, policy makers had in mind environmental matters, mass protests were no longer part of the landscape. Additionally, phrases like 'animal rights' and 'animal welfare' began to appear in everyday language (Yearly 1993).

The last decade of the twentieth century is remembered by environmentalists for the Rio Summit (1992) which, up to that time, was the biggest international conference on environment. The issues discussed concerned global warming, biodiversity, sustainable development, the right to development for the less developed countries and forestry. The Rio Summit gave emphasis to the global character of environmental problems and the need for a global attempt to face such problems, even though attempts at international cooperation had been made before (ibid.).

The first decade of the twenty-first century was inaugurated by the 2002 Earth Summit, the World Summit on Sustainable Development (WSSD), which was held in Johannesburg. The meeting was held to discuss sustainable development and built on the declarations of Stockholm and Rio. According to the declaration of this Summit, the participants agreed on advancing and strengthening the interdependent and mutually reinforcing pillars of sustainable development, economic development, social development and environmental protection at the local, national, regional and global levels. It was decided that the rich diversity of the planet would be used for building constructive partnerships for change and for the achievement of the common goal of sustainable development. Finally, it was agreed that in order for the goals of sustainable development to be achieved, more effective, democratic and accountable international and multilateral institutions are needed.

The 2005 World Summit reaffirmed the commitment 'to achieve the goal of sustainable development, including through the implementation of Agenda 21 and the Johannesburg Plan of Implementation'. The participants agreed on

> undertaking concrete actions and measures at all levels and to enhancing international cooperation, taking into account the Rio principles. These efforts will also promote the integration of the three components of sustainable development – economic development, social development and environmental protection – as interdependent and mutually reinforcing pillars. Poverty eradication, changing unsustainable patterns of production and consumption and protecting and managing the natural resource base of economic and social development are overarching objectives of and essential requirements for sustainable development.

In this framework, environmental groups are organised around concepts such as the one of sustainability. This is what contemporary social movements are all about. Even local protests demand the implementation of policies that protect the environment, while implementing development projects. Such demands coincide with typical sustainable development approaches and concern interests focused on family health, preservation and protection of local ecosystems, and maintenance of local economic resources. Their claims are frequently opposite to claims coming from the state and producers, who favour economic values over environmental protection. They do however agree on the idea of sustainability; the only difference is that they interpret it differently. For example, in the case of the ASG both the bureaucrats who support the bridge project and the members of ASG claim that their proposals aim at protecting the environment of Istanbul which, in their opinion, is unique. This is illustrated by the comments made by one bureaucrat and one ASG participant during the interviews with them. On the one hand a high-ranking bureaucrat from the Ministry of Public Works and Settlement stated: 'To build a road there are certain criteria to be followed. The environment is very important to our lives. This is why we have to plan carefully, putting the environment first.' On the other hand, even less active participants of ASG claim that one of the main reasons for reacting to the bridge project is the environmental effects of such a project. As one of the told me during an interview: 'The environment is the most important thing for human beings.'

Furthermore, the arguments of ASG about the environment follow the rhetoric of the sustainable development discourse in global terms. In the English version of the ASG website is mentioned the following:

> a. Cities such as Stockholm, Barcelona, Florence, Athens, Lisbon, and Oxford are closing their city centres to vehicles that run on gasoline or diesel. More than one hundred cities of the European Union are expected to do the same soon.
> b. In Italy, work sites that employ over 300 employees oblige the employees to share their vehicles when the ride to work.
> c. Another application is the use of electrical or low-emission vehicles where there is heavy air pollution (www.arnavutkoysemtgirismi.tr).

However, the difference is that for the ASG, the protection of the environment means the bridge will not be built at all, whereas for the Ministry of Public Works and Settlement, it means the bridge will be built in Arnavutköy. Environmental protection is an undeniable value. One might even say that the discourse of both sides on the environment becomes instrumental in order to legitimate their stance and that different interests determine different instrumental use of environmental protection (Georgoulas 2006). Yet, whose perception is going to prevail is a question of power relations. As Kousis and Eder (2001: 21) underline, 'The environment is certainly a conflictual discursive field in which sustainability discourses force other discourses to adapt. Nevertheless, whether such dominant discourses assimilate or exclude other discourses depends on the configuration of power.'

Thus, environmentalism is not a question of scientific truth but a question of who has the power to accommodate science with the data in order to conclude what is good and what is bad for the environment. For example, in the discourse about the climate change there are different opinions. In the Academy Award-winning documentary, *An Inconvenient Truth*, the former US Vice President Al Gore claims that, after reviewing scientific data on annual temperature and CO_2 levels in Antarctic ice core samples, it is certain that our planet is in the middle of a 'climate crisis' which people have the power and the duty to interrupt. Responding to these claims Roy Spencer, a principal research scientist at the Earth System Science Center of the University of Alabama in Huntsville, wrote an open letter to Gore criticising his presentation of climate science in the film, asserting that the Arctic had a similar temperature in the 1930s before the mass emissions of carbon dioxide began. Former University of Winnipeg geography professor Dr. Timothy F. Ball rejected Gore's claim that there has been a sharp decrease in the thickness of the Arctic ice cap since 1970, stating that the data was taken only from an isolated area of the Arctic and during a specific cooling period (http://en.wikipedia.org/wiki/An_Inconvenient_Truth).

Even the Kyoto Protocol[2] has been subject to criticisms. As biogeographer Philip Stott (1998: 1) writes about the Kyoto Summit:

> To hear ecologists talking about 'halting' or 'curbing' climate change was deeply disturbing, but for them to try to make the world believe that this 'stability' might be achieved through manipulating just a few variables out of the millions of interlinked and dynamic factors which govern the world's climate is frankly sinister. Let us be blunt; it is a lie, a disgrace to the subject, and a scientific nonsense.

Therefore, in the climate change debate, it seems that science assumes an instrumental role which can be seen as an effort to prevail in the discourse. As Richard S. Lindzen, an atmospheric physicist and the Alfred P. Sloan Professor of Meteorology at MIT, wrote: 'Gore is using a biased presentation to exploit the fears of the public for his own political gain' (ibid.). An example is the debate on the definition of sustainable development among

2 The Kyoto Protocol was introduced on 11 December 1997 during the United Nations Framework Convention on Climate Change, held in Kyoto, Japan. It concerns an amendment to the international treaty on climate change that assigns mandatory gas emission limitations. The aim of the protocol is 'to reduce greenhouse gas emission concentrations in the atmosphere at a level that would prevent dangerous anthropogenic interference with the climate system' (http://unfccc.int/resource/docs/convkp/kpeng.pdf). So far, the protocol has been ratified by 169 countries.

Northern and Southern states (Briassoulis 2000; Eden 2001; Argyrou 2005). As long as Southern nations agree to 'play' the game with the rules posed by the North, which are translated into the discourse on 'sustainable development', the former are doing no more than to contribute to the 'conditions responsible for their inferior position in the global hierarchy ... The better he plays his role as "sustainable developer", the more rounds he loses in a game that is always already lost' (Argyrou 2005: 165).

Similarly, small environmental groups adjust themselves to the culture generated by larger international environmental organisations (Rootes 1999) in order to legitimate themselves as actors in the environmental discourse. As Eden (2001: 81) claims, 'environmental and socio-political movements have also harnessed the power of nature to pull their own particular ploughs'. Therefore, while it is true that environmental concerns become public and cause mobilisation when they directly affect ordinary citizens and the environmental damage done is so conspicuous that it makes both citizens and politicians take interested in the issue, the association of local actions with global bodies and agendas provides the former a legitimate platform to express concerns and opposition.

Environmentalism as a social movement

There are cases of environmentalism which has developed from a form of scientific criticism of a wider social movement (Jamison et al. 1990: 1). Such environmental movements come into conflict with and challenge political and economic power-holders over issues of ecosystem use (Kousis 2004), for example the campaign of *Earth First!* about the Californian redwoods (London 1998), the *Wise Use Movement* in North America, that stresses the neglect of local communities by big environmental organisations (McCarthy 1996) and local environmental protests such as the one discussed in Chapter 4, which articulate environmentalism with issues of ethnic and religious identities (Ignatow 2005a), democracy (Arsel 2003; Öncü and Koçan 2002) and human rights (Sachs 1997; Öncü and Koçan 2002). These campaigns differ from professional organisations because they do not have hierarchical structures but they form a co-operation network. This constitutes the analysis of environmentalism as action which assumes the form of social movements with political references because as Wapner (1996: 1) claims, 'environmental concerns fundamentally involve politics'.

Today the number of environmental organisations existing worldwide is more than 100,000 (Wapner 1996). This number includes not only institutionalised environmental groups but also non-institutionalised environmental movements, which are 'broad networks of people and organisations, engaged in collective action in the pursuit of environmental benefits' (Rootes 1999: 2). Stern et al. (1999: 82) define environmental movements as 'being embedded in a broader society and engaged in struggles in a policy system that includes not only elements of the state but also opponents'.

According to Brulle (2000), environmental degradation is associated with social organisation and this is why the solution for environmental problems is social change. During the 1980s, in the so-called Second World the environmental movement flourished because of the articulation of environmental concerns such as pollution with national sovereignty and public health claims. In the Third World, developmental demands are incorporated into the environmental agenda of social protests, a fact that gives them a radical character (van der Heijden 1999). An example of this can be seen in the *Uttaranchal* movement, a successor of the widely known *Chipko* movement, in the Indian Himalayas (Rangan 1996). It emerged during the 1970s as a reaction to policies that restricted the opportunities for economic development: 'The *Uttaranchal* movement expresses the frustration of some participants who realise that their protests have neither gained wider political support, nor succeeded in forcing the Indian government to negotiate the issue of the creation of a separate state' (ibid.: 206) which is among their demands.

On similar grounds, Rootes (1999) claims that Third World environmental campaigns perceive environmental problems as embedded in other economic, political and cultural issues and they tend to emphasise the importance of political and economic processes in their struggles to protect the environment. His view is supported by Haynes (1999) who states that environmental conflicts in the Third World are articulated with wider demands of socio-political and economic reforms. For example, Moore (1996) discusses a case of conflict in the Kaerezi area of Nyanga

District in eastern Zimbabwe, which at a first glance seems to be centred on access to environmental resources. However, deeper analysis reveals that the struggle was simultaneously about cultural meanings of the environment and gender relations. Within this framework, it is more appropriate to speak about transenvironmental movements rather than environmental ones; that is, movements that relate environmental crises to social crises. Finally, it is also fair to claim that environmental conditions are the result of political choices.

The study of ASG is a discussion of such a collective action; a transenvironmental collective action in the context of which the environment acquires certain cultural and political connotations. As discussed in Chapter 4 this fusion is not unique in the case of ASG. The Similarities that Arnavutköy District Initiative shares with other campaigns in terms of talking about the environment in tandem with other social issues, illustrate that in Turkey there is a social movement in terms of a collectivity of people united by common beliefs in accord with central values, such as democracy, human rights and environmental conservation (Nash 2005), and a determination to challenge the existing order in pursuit of these beliefs (Schönwälder 2002).

This is where the ASG draws its features as a social movement organisation – on the similarities which it shares with other local campaigns which, together, constitute a social movement. In a declaration sent to be published in the website of the Mega Revma Fellowship – an association of Orthodox Christian ex-residents of Arnavutköy – it is mentioned that the residents of Arnavutköy 'encouraged by the fact that in the Aegean town of Bergama, ordinary citizens have recently managed to stop a gold production project that risked leaking cyanide into their water table have initiated their own struggle' (www.megarevma.net). Furthermore, the support by national (Çekul Vakfi, DHKD, Chamber of Architects, BSNN) and foreign NGOs (UNESCO, EUROMEDCITY, Fellowship of Mega Revma) as well as with other neighbourhood associations like the Kuzguncuk Association illustrate the bonds between ASG and all these other groups. The added participation of individual activists and famous personalities of the country creates a framework in which Arnavutköy becomes everyone's backyard, and the ASG struggle becomes part of the national and international struggles. These issues are going to be discussed in the following section.

ASG in the landscape of social movements studies

Theories of collective action can be traced back to the structural–functional sociological tradition of Talcot Parsons. Obviously influenced by Durkheim, Parsons wrote about the revolutionary movements of the 1920s and 1930s, analysing them under the prism of anomy. Robert Merton went further and claimed that such actions are not only deviant but also mechanisms of adaptation to new forms of organisation (Psimitis 2006). A completely different argumentation was developed by the Chicago School – the symbolic interaction approach which draws a distinction between social organisation and collective behaviour. Movements were fallen into the second category and along with other collective action manifestations were seen as mechanisms of positive social transformation (Edelman 2001). Marx and Engels (1998 [1913]) wrote about collective action in relation to the system of production and viewed conflict in capitalist societies as revolving around the fundamental contradiction between the bourgeoisie and the proletariat (Edelman 2001: 288). This approach shed light on class struggles and examined the collective actions of the workers against those who possessed the means of production. Finally, theorists of the Rational Choice School (Olson 1965) claimed that collective action derives from private motives alone. These first theories of collective action gave food for thought to what was later called theories of social movements. These theories can be categorised into two main groups with a number of subgroups each (Psimitis 2006).

The first group, known as the New Social Movements (NSM) theories, concern the European tradition on the study of social movements. Some of the most influential proponents of this approach are Jürgen Habermas, Alain Touraine, Alberto Melucci, Ernesto Laclau, Chantal Mouffe and Claus Offe. New Social Movements theories emerged and evolved in

Europe during the 1970s as a response to the Marxist analysis of collective action. Although the proponents of this approach have focused on different aspects of social movements, they agree that social movements result mainly from the symbolic realm of social relationships and not from class struggle as the Marxist approach claimed; that is, NSM theorists focus on movements that seek to change symbolic and cultural values within civil society in confrontation with the state. According to Alberto Melucci, NSM in complex societies shifted the field of struggle to the symbolic realm. This is exactly why their actions are political: 'Even though they do not explicitly target against political and economic structures, they are agents of change with clear political and economic implications. Movements such as the feminist, the student, the environmental and the peace movement – even though they act in the symbolic level – they have successfully intervene in policy making' (Melucci in Psimitis 2006: 161).

However, the control over symbolic means of political pressure can only bring about symbolic responses by policy makers (Ho Kwok-Leung 2000), and this is why the second set of theories centres around how contemporary social movements mobilise in order to achieve their goals. This set of theories draws on the American tradition of Resource Mobilisation Theory (RMT) and it is sometimes seen as a strategy-oriented theory (Edelman 2001). For Resource Mobilisation Theorists, contemporary social movements are forms of collective action with political agendas and appropriate means to put them in effect. In this process, an important role is assigned to the resources that a movement can mobilise such as time, numbers, money, powerful third parties, ideology and leadership (Şimşek 2004). The attainment of the goals a movement sets is related to the ability to acquire appropriate resources (Psimitis 2006). Characteristic proponents of this theoretical trend are Charles Tilly, Donatella Della Porta and Mario Diani as well as Mayer Zald, John McCartly and Doug McAdam.

According to Psimitis (2006), RMT focuses on social movements as rational collective actions that use available resources to accomplish their goals. The contribution of RMT to the study of social movements lies in the fact that it distinguishes between social movements and social movement organisations. A social movement organisation can be a small group that identifies its goals with the goals set by wider and pre-existing social movements and thus becomes part of them. This distinction helps the empirical study of social movements by creating analytical tools for differentiated and different-scale actions. Finally, an RMT approach of social movements as political actions reveals a less romantic aspect of them which illustrates their political aspect. As Şimşek (2004: 117) puts it, 'the success of a group is measured by whether or not it is recognised as a political actor, and whether or not any increase is observed in material benefits'.

Resource Mobilisation Theory discusses individual participation in collective action mainly when personal interest is involved. In this sense, the symbolic aspects of individual action related to the formation of collective identities without cost-effective calculations are completely disregarded and material interests underscore cultural identities (Psimitis 2006). If this is the case, it is difficult to explain how 'social movements, usually of the very poor, emerge with few resources or where overt organisation in contexts of extreme inequality, severe repression, and hopeless odds endangered participants, producing "shadowy" (Piven and Cloward 1977), "submerged" (Melucci 1989), or "hidden" forms of resistance that might or might not lead to collective action (Burdick 1998)' (Edelman 2001: 290). According to Psimitis (2006), another pitfall of RMT is that giving a central place to 'interest' in the analysis of social movements may confuse this 'interest' with other forms of collective actions such as interest groups. In cases like the one discussed thoroughly in this book, the term 'interest' risks being confused with NIMBYism, which – as discussed in Chapter 3 – is not a plausible speculation for ASG.

From another view, New Social Movements theories tend to disregard economic and power inequalities and it seems that these differences do indeed matter particularly in non-western developing countries where the welfare state is weaker than in Western settings. Thus, students of social movements of such countries refer to popular movements rather than adopting the term new social movements (Edelman 2001). Furthermore, some claim that NSM theorists neither examine how social movements are organised and function nor assess the importance of their role in political change (Psimitis 2006). Finally, a point that seems more relevant to anthropologists is that the underlying assumptions of these theories have yet to be tested in micro-level analysis (ibid.), a fact that makes NSM theories more vulnerable to critique than RMT from an anthropological point of view.

Globalisation and social movements: The view from anthropology

As mentioned above, both New Social Movements theories and Resource Mobilisation Theory have been subject to critical assessment that illustrated each paradigm's drawbacks. As Psimitis (2006) claims, it is obvious that the weakness of NSM theory is the strength of RMT and vice versa. In the meantime, a new paradigm has been emerging which combines elements from both the American and European tradition in the study of social movements. I call this new group of theories the Global Social Movements Theory (GSMT), with proponents such as Arjun Appadurai (1996), Hillary Cunningham (1999), Pamela Martin (2003), Marc Williams (2003) and June Nash (2005). These approaches focus on 'the tensions created by globalisation processes and the ways in which populations affected by them, re-envision their positions as they respond to global tendencies in culturally and historically distinct ways' (Nash 2005: 13). Even though Appadurai (2000: 15) states that 'the sociology of these emergent social forms part movements, part networks, part organisations has yet to be developed', a number of studies have flourished in the social sciences concerning such issues as indigenous rights protests (Findji 1992; Warren and Jackson 2002; Martin 2003), housing campaigns (van der Linden 1999; Ho Kwok-Leung 2000), urban poor struggles (Bennett 1992; Canel 1992), protest against infrastructure developments (Hall 1994; Voulvouli 2004a, 2004b; Doane 2005), environmental protests (Carcía 1992; Öncü and Koçan 2001; Arsel 2003) and Latin American social movements (Escobar and Alvarez 1992; Escobar 1996; Peet and Watts 1996; Rangan 1996; Warren and Jackson 2002). The problem until recently has been that 'anthropologists on the one hand and sociologists and political scientists on the other have had little impact on or awareness of each other's efforts to understand social movements' (Edelman 2001: 286).

GSMT manages to combine the 'grand theory' preferences of sociologists and political scientists with the micro-level approach favoured by anthropologists. Of course, anthropologists have been studying local collective action long before globalisation theories came into theoretical mainstream. Marginalised groups have always been the subject of anthropology either outside of or within the developed world as well as their decision to protest against dominant structures. As Nash (2005: 177) puts it: 'anthropologists are by inclination and profession predisposed to study the peripheral phenomena of everyday life everywhere in the world and especially in marginal areas'. For example, the reaction of the 'people without history' – to use Eric Wolf's words – to the newcomers of the old world has been studied by anthropologists since the late nineteenth century.

In this sense, GSMT is a set of approaches that has followed anthropological perspectives in tandem with macro-theories, focusing primarily on the effects of globalisation on different groups of people. The similarity of GSMT to New Social Movements theories lies in an emphasis on the identity of these groups of people, although the focus of these studies is on how these identities emerge and evolve through the mobilisation actions, and not so much on the structural influence that identity has on the mobilisation (Appadurai 1996; Nash 2005). For example, June Nash (2005) explores how the Mayas of Mexico and Guatemala react to processes of deterritorialisation and resettlement forced by globalisation and explains how concepts of locality and belonging to a particular place evolve as the reaction of the residents rises. The similarity that GSMT approaches share with Resource Mobilisation Theorists is a focus on the ways in which these groups mobilise because it reveals the different reaction each group has to global effects (Appadurai 1996; Escobar 1996) which in turn reveals the cultural reasons for resisting (Theodossopoulos 1997, 2000, 2002). A case study of resistance from Greece explores the historical and cultural reasons motivating local reactions to the governmental conservation efforts, reinforced by global environmental notions. The study describes the case where the residents of a Greek island protested against the decision of the Greek government under the pressure of the EU and environmental conservationists, to ban the use of an area that housed the careta-careta turtle, an animal threatened by extinction. As Theodossopoulos claims, the cultural reasons behind the objections of the local population to the government conservation programme had to do with taking care of one's household. The villagers' resistance stemmed from concerns related to the welfare of their households and in giving priority to this concern, 'they fulfilled their public role as responsible members of their households and their community' (Theodossopoulos 1997: 264).

The Greek case study illustrates how globalisation in terms of globalised ideas on environmental protection reinforces simultaneously national politics, which in turn mobilise identity politics (Theodossopoulos 1997, 2000, 2002). As Theodossopoulos describes, the Greek government was forced to proclaim the area as a national park not only by environmental groups, which had access to the media and thus to public opinion, but also due to EU regulations on preserved areas. Given all these factors, the Greek government did not have any alternative at the time, even though the political cost in this area would have been steep.

The resistance that Theodossopoulos describes is a protest against transnational forces and shares similarities with other resistances in other parts of the world which also struggle against globally disseminated values (Fairhead and Leach 1996; Burnham 2000). Communication technology has made people aware of these struggles and therefore local groups have the chance to discover that somewhere on the planet there are other people with the same concerns and problems (Appadurai 1996). More than twenty years ago, Benedict Anderson (1983) defined the process of the awareness of sameness in the creation of imagined communities. In the new millennium, this process is even more present, as nowadays we are dealing with 'the intensification of worldwide social relations which link distant localities in such a way that local happenings are shaped by events occurring many miles away and vice versa' (Kearny 1995: 547). In such a reality the notion of space is understood socially rather than territorially (Williams 2003); the formation of identities is transnational (Appadurai 1996; Cunningham 1999) and, therefore, the construction of alliances is in turn transnational as well. As Kearny (1995) claims, local groups situate themselves in globally informed networks, re-define their projects and obtain allies from other parts of the world.

'Third Bridge: Hayir, No Nein, Non': The transnational character of local social movements

World System Theory asserts that the global perception of the world has been brought about mainly by economic capitalism. Today 'global corporations seek to insinuate brand concepts not only in market fields but also into the broader culture thus reshaping communities into newly privatised pseudo-public spaces' (Edelman 2001: 308). While the market is promoted as an ideology by the state that seeks to attract foreign capital in a highly competitive global market place (Williams 2003), state policies evolve at the foundations of the globalisation process (Durrenberger 2003: 275) in terms of flexible global capitalisation that creates deterritorialisation, hybridisation and fragmentation (Nash 2005), issues that mobilise local-level reactions (Appadurai 1996). These have resulted in the domination of globalisation discourse in social sciences the last twenty years.

As Ahmet Öncü and Gürcan Koçan (2001) write, globalisation is characterised by the possibilities of the market in a worldwide level and thus, the political and cultural social process of globalisation must be seen in relation to the logic these possibilities reinforce. This means that supranational economic powers dominate in the arena of social rights, which in turn results in pressure from citizens whose rights are being overlooked. These citizens organise themselves in groups with a transnational character since the forces they oppose to are also transnational. In other words, globalisation has contributed to the increase of the power of regional and supranational groups, whereas the impact of the nation-states has begun to diminish. Members of groups which suffer in their local regions call for support from supranational powers such as the EU. In the Turkish context, examples such as the Kurdish movement and environmental groups confirm this tactic (Şimşek 2004). The Bergama, the Ilisu Dam and the Tunceli cases are vivid examples of such interactions. The ASG has also tried and to a large extent has managed to create alliances with supranational organisations such as the UNESCO and other groups such as the Mediterranean Cities Organisation.

As Lewellen (2003) claims, globalisation has a two-fold effect: On the one hand it strengthens global economic institutions, such as the World Bank and the International Monetary Fund, and, on the other it gives rise to grass-roots activities which mobilise against the constraints on them due to economic globalisation. Similarly, Ignatow (2008a: 2) argues that 'globalizing processes not only engender global uniformity but also animate ethnic and religious identities and give rise to hybrid forms of culture and politics'. Furthermore, Williams (2003) claims that globalisation technology gives rise to organisations focusing on global issues, such as environment and human rights. For example, environmental politics have been shaped by institutional forms of coordinating collective actors that maintain relationships with transnational actors. In this way they create transnational fields of political action and policy-making (Kousis and Eder 2001). As a result, in recent years, a number of groups have become transnational. They are organised across state boundaries and work toward environmental protection at global level. Some of these groups enjoy the membership of millions of people worldwide, which is one of the factors that gives them their transnational character, and their budgets are bigger both than the environmental budgets of most countries and than the amount the United Nations Environment Programme (UNEP) spends annually on its operation costs. Examples of such organisations are Greenpeace, Friends of the Earth (FoE), Conservation International and the World Wildlife Fund (WWF) (Wapner 1996).

Lately anthropologists who study social movements have realised that, in the *global ecumene*[3] we live in, focusing just on 'fragments of culture'[4] is sometimes not enough.

3 The *global ecumene* is the interconnectedness of the world by way of interactions, exchanges, and related developments, affecting not least the organisation of culture (Hannerz 1996: 7).

4 This expression is borrowed by the title of the book edited by Deniz Kandiyoti and Ayşe Saktanber, *Fragments of Culture: The Everyday of Modern Turkey* (2002).

> Anthropologists who once ignored the intrusion of national and international in their field sites are now among the principal observers of social movements, particularly those of indigenous people, women, and the disinherited, as they seek a new relationship with the states in which they are included. Although the potential of these movements are often underestimated, it is in these circles once considered marginal to global processes that major transforming changes have been occurring. Because of their cultivated peripheral vision, anthropologists are in a position to assess new directions (Nash 2005: 22).

One of the first anthropological works on social movements under a global prism was that of Arjun Appadurai. In his 1996 book entitled *Modernity at Large: Cultural Dimensions of Globalisation* Appadurai maintains that contemporary public spheres have a diasporic character, created by mass migration and electronic mediation, which transcends the borders of nation-states. As he puts it, 'electronic mediation and mass migration mark the world of the present not as technically new forces but as ones that seem to impel (and sometimes compel) the work of imagination' (1996: 4) and imagination, in its collective forms, can become the fuel for action (ibid.: 7) through the creation of micronarratives[5] that are in turn the fuel of oppositional movements (ibid.). He calls these movements culturalist, after having defined the term culturalism as 'identity politics mobilised at the level of the nation-state' (ibid.: 15). Moreover, he suggests examining them as movements that represent the 'cultural dimensions of globalisation' and thus they represent difference in terms of identities mobilised by cultural differences. Put differently, his concern is the 'articulation between culture and group identity' (ibid.: 13) in the context of globalisation, in terms of larger national and transnational politics.

The networks created through processes of mobilisation of identities (for example) might not be global in the sense that they have representatives across borders, but they are global in consciousness. Their alliances can be better described as transnational a term which 'signifies that many

5 The term micronarratives is intended as the cultural products such as film, television, music and other expressive forms, which allow modernity to be re-written more as vernacular globalisation and less as a concession to large-scale national and international policies (Appadurai 1996: 9).

of the linkages in question are not "international" in the strict sense of involving nations – actually, states – as corporate actors; the actors may now be individuals, groups, movements, business enterprises, and in no small part it is this diversity of organisations that we need to consider' (Hannerz 1996: 6). 'Transnationalism' also signifies the alliances created between some nations but not in the whole world (Kearny 1995). After all, as Hannerz (1996) claims, there are nations that do not take part in globalisation but even those who do take part are weakened by the increasing power that global and local organisations assume.

A campaign from Turkey, which has not been mentioned earlier in Chapter 4 but is a fine example of this claim, is the Akkuyu resistance. It revolves around the construction of a nuclear power plant. The campaign has made its struggle public both at the national and the international level. The opposition began in 1976 when the area was selected as the site of a nuclear plant. However, the protest became stronger in 1994 when Turkish environmentalists were larger in numbers, and after the Chernobyl disaster had made people more aware of the risks they would run if a nuclear energy plant were built in their area. Organised protest began in 1994 when citizens from Akkuyu demonstrated outside the office of the then Prime Minister, Tansu Çiller. This demonstration resulted in a meeting between the Prime Minister and the activists who proclaimed that 'the Mediterranean will remain blue'. The nuclear plant project was solidly rejected by the villagers of the area, a stance that became visible in the results of an unofficial referendum held in 1999, with 84 per cent of the residents of the area voting against it. The Akkuyu protest assumed various forms: nationwide demonstrations, channelling international activists and educational campaigns against nuclear power through Greenpeace. Those actions brought results, such as the international support from Europe and North America, the endorsement of other international environmental organisations such as the International Physicians for the Prevention of Nuclear War (IPPNW) and the Canadian Coalition for Nuclear Responsibility as well as the formation of Turkish Cypriots Coalition Against Akkuyu and the initiation of a campaign by Greek Cypriot activists under the banner of the Green Party. Finally, Greek environmentalists expressed their support and through their representatives brought the debate to the European parliament.

In this way, 'local concerns come to be identified with global issues and local campaigns with the global phenomenon and so are seen to be in everyone's backyard' (ibid.: 298). This is why radical environmental conflicts are rarely local in essence. They may start as a local protest but they almost always find themselves at a national and frequently transnational level. In the case of ASG this is conspicuous from the language that ASG uses, which is in accordance with the rhetoric of global environmentalism. Besides, ASG has never tried to conceal the fact that it appeals to the sentiments of the international community for the protection of Arnavutköy. As shown in Figures 6 and 7 this effort is symbolically reflected on their banners, some of which are written in languages other than Turkish.

A recent work from the anthropology of social movements – through a global prism – was published in 2005 as an edited volume by June Nash, entitled *Social Movements: An Anthropological Reader*. In this volume, anthropologists discuss social movements through the prism of globalisation theories and relate them to issues such as civil society, religion, deterritorialisation, privatisation, individualisation and global cosmopolitanism. All these issues are examined diachronically as they evolve through the struggles of movements against what is perceived to be threats to central values. One of the most important issues that Nash raises in the introduction is that, as the case studies in the volume document, the efforts of social movements to make sure that social justice is not overlooked as globalisation expands. This means that any collective effort that is guided by principles of social justice falls under the same category. For example, as she claims, 'feminism is a movement against sexism that has expanded to include men and gays as well as women. The goals of many feminist groups are the liberation of society from behaviours that constrict the humanity of any one group' (ibid.: 13).

Similarly, Edelman (2001) claims that anti-corporate activists around the world do not only focus on particular problems but link those problems with broader issues of democracy, culture and environment. He goes on to state that ethnographers of social movements should realise that the classical ethnographic approaches are hardly sufficient for studying the deep historical roots or wide geographical connections of most contemporary mobilisations. Thus, on the one hand, local actions should be looked at as part of wider networks not only in terms of space but also in terms of scope. For example, an environmental group can be part not only of transnational environmental movements, but also of human rights movements, movements fighting for democratisation and sexual emancipation movements. In Turkey there are groups which constitute wider movements and at the same time they are being constituted by these movements. Economic and media globalisation has given rise to imagination (Appadurai 1996), which in turn has given rise to movements that have a hybrid character, namely religious–environmental, ethnic–environmental and women's–environmental (Ignatow 2005a, 2008a). On the other hand, 'there are significant aspects of social movements such as "submerged" organising, informal networks, protest activities, ideological differences, public claim-making, fear and repression and internal tensions that raise questions that can be addressed only through ethnographic or ethnographically informed historical research' (Edelman 2001: 310). Such a study is the analysis of Stephen (2005) about the identity politics of women participating in the grass-roots movement of CO-MADRES in El Salvador. As she claims, in an effort to project a sense of 'sameness' to the outsiders the 'participation (in such movements) is a constant process of negotiating difference' (ibid.: 67) a fact which create tensions within the group. Another case in which internal processes of movements are visible to the ethnographic eye, is Kasmir's study (2005) about the union of an automobile plant in Tennessee, USA, which examines how class identity is shaped in the course of social action as well as the character and the conditions in which different identities are made.

All these groups described above share similarities with the ASG. Even though they were born out of diverse social circumstance and struggle against different issues, they all go beyond the subject that triggered their formation and somehow they put forward demands related with democratic participation, human rights, social justice. Furthermore, they all centre on common identities which emerge and evolve by way of the protest. In this context, it is worth mentioning the phenomenon of globally informed but transnationally active social movements. Globalisation has brought about common imagination which in turn has brought about transnational alliances. Contemporary social movements even in their most local expression are integrated in global discourses and are part of wider social movements, particular expressions of which exist around the world, namely ethnic, feminist, human rights and environmentalist. As some observers would claim, contemporary social movements are part of the global 'civil society'.

While it is true that today nation-states are key actors in addressing issues such as the environment, cultural heritage and human rights and that activist efforts play a major role in pressuring them, there has to be acknowledged that there are other arenas of politics. Even though these arenas are separate from the state and are situated in the so-called global civil society, they intermingle with the state through the practice of 'world civic politics'. This term describes the effort of NGOs to engage in world politics in a way that results in the politicisation of global civil society (Wapner 1996). Therefore, collective experience is penetrated not only by state institutions but also by other spheres that influence the character of public activity. Such a sphere is the economic realm, for example, that shapes many people's concerns and involvement to public life. This occurs both within countries and beyond national borders – anywhere that the existence of mechanisms, in the form of social and cultural networks, shape collective behaviour. Such networks include voluntary associations in reference to science, trade, cultural expression, religion and production. Civil society, as the realm existing above the individual and below the state, is based on friendship, custom, the market and voluntary affiliation. In this sense, it consists of a network of economic, cultural and social practices which recently has begun to make sense on a global level (ibid.). Thus, the so-called global civil society is 'that slice of associational life that exists above the individual and below the state but also across national boundaries' (ibid.: 4).

The increasing number of NGOs across states and the development of transnational network in combination with the internationalisation of national economies as well as the weakening of national sovereignty caused by globalisation and rapid technological change, gave rise to the emergence of non-state communities (Williams and Ford 1999). National identities begin to give their place to global identities shaped by communities of women, workers, refugees or peasants. Such forms of identity do not necessarily have as reference point the global region but they can be transnational as much as regional and global (ibid.).[6] For example, Toth (2005) discusses how an Islamist movement initially emerged in Egypt assumed global character and identity after the September 11 attacks on the World Trade Center in New York. In this context, civil society actors link civil society across boundaries and in this way they shape and re-shape international politics to the extent that these actors assume the character of a systematic collectivity; a social movement. Hence 'social movements challenge the constitutive practices of the international system and in particular expose their potential for offering alternatives and for instigating far-reaching social change' (Williams and Ford 1999: 287).

Following this, social movements' demands can be seen as claims for power in order to fulfil the goals of certain groups of people. Such movements include the environmental movements. For example, McCormick (1995: xv) claims that 'environmentalism is defined as the promotion of values, attitudes and policies aimed at reaching an accommodation between human needs and the limits of the natural environment'. The sooner this aim is reached, the better the future for the planet and its inhabitants will be (Cotgrove 1976). In this sense, environmental movements demand action; that is, new policies, which focus on the realisation of this aim.

Civil society and the state in Turkey

In Turkey the period that followed the *coup d'état* of 1980 created new dynamics between the state and society that gave birth to civil actors of neo-liberal politics such as commerce chambers and business associations like TÜSIAD (Turkish Industrialists' and Businessmen's Association) and

6 See also Appadurai 1996.

MÜSIAD (Independent Industrialists' and Businessmen's Association). Some claim that the state has used these associations in order to implement neo-liberal economic and political measures and thus their role was hardly civil, in terms of non-economic and non-political. On the contrary, they were rather important actors in policy making. This implies that the development of civil society is associated with the political culture within which it emerges and in Turkey, the political culture is embedded into a highly centralised state infrastructure (Beşpınar-Ekici and Gökalp 2006).

Both the examples of the environmental associations and the environmental campaigns discussed in Chapter 4 confirm this aspect. As discussed in the chapter environmental organisations in Turkey operate either in parallel or against the state, the latter being a solid point of reference. This is also evident in the case of the ASG that not only holds the state responsible for the negative development that the country has followed but also its whole campaign focuses on rhetoric and claims directed to and against the Turkish state. More importantly the identities of ASG participants as they emerge and evolve in the context of the initiative have also the state as point of reference. Secularism, leftism and religion, as argued in Chapter 3, are identities which in Turkey are constantly filtered through statist influence. Consisted with this view, I claim that the mainstream notion of 'civil society' not only risks overlooking the differences between the ASG and other groups acting in Turkey simply because they are considered parts of the same milieu (see Chapter 4), but also overshadows the political and socio-cultural peculiarities of the group.

How civil is 'civil society'?

The concept of imagination, as described by Appadurai (1996: 8) as 'property of collectives and not merely as a faculty of the gifted individual that is the fuel of collective action', brings to mind the concept of fantasy as discussed by Navaro-Yashin (2002: 4), who claims that

> the critical capacities employed in 'ideology' or 'discourse' critique are not only the prerogative of trained intellectuals [...] In Turkey during the 1990s people from all

> sections of society were constantly involved in criticizing various manifestations of the state in the most sophisticated manner [...] And yet, simultaneous practices of re-production, regeneration, and re-reification keep re-dressing 'the state' in a variety of garbs [...] For the very people who critique the state also reproduce it through their 'fantasies' for the state.

The combination of these two concepts, imagination and fantasy, can be very useful in the study of local mobilisations in Turkey. Using the concept of imagination as fuel for action as described by Appadurai, one can illustrate how the character of most of local reactions in Turkey, are against decisions made by the state, while the concept of fantasy as discussed by Navaro-Yashin illustrates the relationship of such actions with what they are opposing; that is, with the state. In this way, 'the state and the people can be examined not as separate but as the same domain' (ibid.: 2), rendering thus the notion of civil society, in the context of which such actions are often situated, problematic.

Civil society as an idea has its roots in the Middle Ages. The possession of property gave rise to political rights and obligations that followed the idea of good governance since the late sixteenth century and especially after the rise of the powers of the state (Mardin 1995). The term 'civil society' was first coined by the German philosopher Georg Wilhelm Friedrich Hegel in his work *Elements of the Philosophy of Right* in 1827. According to Hegel, civil society was the realm existing through dialectical relationship between the state and the family. Ever since, the concept has been subject to different interpretations. Today, the terms 'civil society sector' and 'civil society organisation' refer to a wide range of private organisations outside the state sphere which are neither primarily commercial nor profitable (Salamon et al. 2003). Accordingly, some believe that civil society draws its legitimacy and power from outside of the state sphere. They claim that civil society does not need the state but it is itself a point of reference to liberty (Dahrendorf 1990; Keyman and Içduygu 2003). According to Salamon et al. (2003: 2), 'the rise of the civil society sector may, in fact, prove to be as significant a development of the late twentieth and early twenty-first centuries as the rise of the nation-state was of the late nineteenth and early twentieth centuries', as it is able to counterbalance the statist dominance (Gellner 1994).

Other social scientists stress that the idea of a dichotomy between state and civil society is a problematic characterisation because it does not only forget the fact that civil society needs the state to be protected and to ensure basic social conditions (Hall 1995: 16), but also that civil society does not escape the hegemonic structures that the state and other dominant groups impose upon it (Abélès in Gledhill 1994). As de Tocqueville claimed in 1840, in democratic nations there are conflicts between administration and civil society. Nevertheless, people tend to consider the state as the only force that has the interest and means to defend its citizens from anarchy and insecurity. Thus, they entrust the state with their collective interests as it seems to be their only constant representative. In this way, however, the state defends not only the citizens' interests but also its dominance to public consciousness (in Psimitis 2006). Therefore, the power of the state grows to become organic in a society in a subtle way and becomes part of a series of societal controls and 'normalizing' influences (Foucault 1980). This explains why, in cases of conflicts, clusters of civil society use signifiers of the state in order to legitimate their opposition to it. For in public consciousness, the dominant way to stand up for a cause is the state's way. At the same time, while doing it, people silently legitimate the state's authority, and it is this silent legitimation that makes the distinction between state and civil society challenging.

According to the London School of Economics Centre for Civil Society's (LSE CCS) working definition,

> even if in theory the institutional forms civil society takes are distinct from those of the state, family and market, in practice, the boundaries between state, civil society, family and market are often complex, blurred and negotiable. Civil society commonly embraces a diversity of spaces, actors and institutional forms, varying in their degree of formality, autonomy and power. Civil societies are often populated by organisations such as registered charities, development non-governmental organisations, community groups, women's organisations, faith-based organisations, professional associations, trade unions, self-help groups, social movements, business associations, coalitions and advocacy groups (LSE CCS).

Besides, as Wapner (1996: 153) reminds us, 'actions taking place in the civil society realm or in the state realm have consequences for each other'.

Non-governmental organisations, which are considered as constitutive parts of civil society influence state policies and, by doing that, they create a collective culture. Hence, to understand world events one should see them under the prism of the combination of the state and civil society. However, Neo-Gramscian perspectives claim that civil society and public opinion are defined by those in power through the use of propaganda and the media. In this framework, the study of civil society is linked with the study of politics; that is, 'the study of the processes involved in determining and implementing public goals and in the differential achievement and use of power by the members of the group concerned with these goals' (Swartz et al. in Lewellen 2003: 85).

Therefore, the binary opposition between the state and civil society is an oversimplification for various reasons. First, because civil society organisations (CSOs) – as will be discussed below – use state methods to achieve their goals. Secondly, certain aspects of CSOs illustrate that they actually function like governmental organisations and besides, they are funded by governments. Many CSOs receive state or intra-state funding; for example, technical innovation organisations are often funded by governments (Clark 1995). Thirdly and in reference to this second point, there are the CSOs which are market-oriented and sell their services to aid donors and government agencies to implement projects and programmes (Robinson 1997: 59). According to Edwards and Hume (1997: 276), non-governmental organisations have expanded in number and scale through donor support. As Salamon et al. (2003) claim, civil society seems to be a major economic force. These scholars' research, which was carried out in thirty-five developed and developing countries, illustrates that the civil society sector aggregates expenditures of US $1.3 trillion and is a major employer with a total work force equivalent to 39.5 million full-time workers. As they claim, if the civil society sector was a separate national economy, it would make the world's seventh largest economy.

Civil society in Turkey: 'More as a slogan than as a reality'

The emergence and development of civil society in Turkey has been discussed widely by social scientists (Gellner 1994; Göle 1994; Jalali 2002; Navaro–Yashin 2002; Keyman and Içduygu 2003; Şimşek 2004; Kadioğlu 2005; Rumelili 2005; Beşpınar-Ekici and Gökalp 2006; Gürbey 2006). One of the central issues in the literature focuses on the development of civil society in the post-1980s era, following the end of the coup when Turkish civil society became more visible and vocal – when scholars, intellectuals and commentators influenced by global popularisation, began to interpret even the smallest developments in civil society. For example, the electoral victory of the Motherland Party (ANAP) over the National Democracy Party (MDP), openly supported by the Military Council in the 1980s, was seen as a victory of civil society over the military (Şimşek 2004). In addition, a series of civil-society organisations (CSOs) were founded by Islamic organisations, business associations, Kurdish, Alevi, environmental and feminist groups (Kubicek 2005; Ignatow 2008a). Rhetoric about civil rights, democratisation and citizenship, which belongs to the language of civil-society actors, began to appear in public utterances as civil society became stronger through 'the legitimacy crisis of the strong-state tradition'. The country's tradition of the state as 'the primary context of politics' and central in the production of cultural life was challenged by new actors, new mentalities and the new language of modernisation, as well as democracy as a global point of reference in politics (Kubicek 2005).

For some, these changes were the result of the democratisation programme and unification agenda of the European Union. Their importance stems from 'the changing meaning of modernity' and 'the emergence of alternative modernities' in the Turkish context. Secular mentality, for example, no longer appears to be the only facet of modernity. Both political Islam and Islamist identity have gained ground, in terms of what the current meaning of modernisation should entail (Keyman and Içduygu 2003). Others claim that, after the 1980s, civil society in Turkey was formed on the basis of repression of the associations organising around leftist politics, which gave rise to the development of identity politics as a substitute for the lack of political agendas (Beşpınar-Ekici and Gökalp 2006).

A series of particularistic identities overlooked until then, due to the highly nationalistic environment after the foundation of the republic, came to the scene. In this sense, it is the state that defines civil society and that has the final word as to which organisation belongs to civil society and which does not. This is why local mobilisations have rarely been successful for the protestors. Mobilisations about ethnic and religious identity politics are conceived as threats for the national security and their demands are hardly accommodated. Finally, civil actions with political demands are seen as inappropriate as the civil sphere in the eyes of the state can only articulate demands with apolitical content (Beşpınar-Ekici and Gökalp 2006).

The depoliticisation of the 1980s was followed by the rise of a new apolitical form of civic associations and private media which led to the 2004 Law on Associations, which regulates state's intervention in civil-society activities and allows the collaboration of NGOs with political organisations. According to some scholars, those developments advocated democratisation and growth of civil society (Robins 1996; Göle in Robins 1996: 72; Keyman and Içduygu 2003). CSOs have been considered democratisation agents, mediators between Turkey and the EU as well as actors working towards the modernising and the liberalising the Turkish state, in terms of social accountability. In this climate, even amid military intervention, the discourse of civil society was normatively supported and actively promoted in academic and public life in Turkey during the 1990s. The failure of the Turkish state to respond quickly to crisis situations such as the 1999 earthquakes, and the parallel mobilisation of civil society organisations, strengthened the latter's position in public opinion which gave rise to demands regarding participatory political culture (Keyman and Içduygu 2003).

Nevertheless, other scholars claim that, even though the 61,000 newly formed associations constitutes an astounding number, their impact on public life has been less significant (Şimşek 2004; Beşpınar-Ekici and Gökalp 2006): Civil society in Turkey exists 'more as a slogan than as a reality' (Kubicek 2005). Civil society in Turkey seems weak, passive and controlled by the state, a fact that has its roots in the Ottoman political culture, the bureaucratic nature of the early Turkish Republic and the dependence of Turks by the *devlet baba* ('father-state') (ibid.). The Turkish

Republic followed the social conventions of the Ottoman Empire regarding the restriction of individual participation in the decision-making process, hence a heritage of rigid state–society polarisation (Özdemir 2003).

In addition, the contribution of civil society to democracy has been challenged by scholars like Şimşek (2004), who claims that certain characteristics of civil society in Turkey decisively contribute to democratic aspirations such as the large number of existing NGOs. Even though a more liberal Law on Associations (2004) reduces the interference of the state in NGO activities, removes restrictions on contact with foreign NGOs, and allows NGOs to work with more overtly political organisations (Kubicek 2005), they are not entirely autonomous from the state and military power. Their participation in decision-making is very limited, since about 95 per cent of Turkish NGOs are subject to the earlier *Dernekler Kanunu* (the Law of Associations), which prevents them from participating in politics and cooperating with political parties. This along with the prohibition of civil servants to join political parties and engage in politics render Turkish civil society weak and only when full or even relative autonomy from state power and de-fragmentation of NGO visions is achieved, it can play an important role to democratisation.

Beşpınar-Ekici and Gökalp (2006) argue that in order to have an efficient democratisation process in Turkey, a strong civil society with political agency, negotiating with the state based on local communities' demands, is needed. They claim that today, civil society in Turkey is ineffective in policy making because it is not legitimate in the eyes of the state. However, the dependence of civil society from the Turkish state seems less strong in cases of CSOs that focus on the EU agenda and other international funders, such as the World Bank. The rise of market channels rendered CSOs autonomous and the influence of globalisation in terms of intensification of economic, political and cultural relations across borders as well as the impact of the process of European integration were decisive factors for the character of civil society in Turkey (Kubicek 2005). In this process, the issues of citizenship and civil society as mentioned in the Accession Partnership Document (APD) issued in 2000, are among the non-negotiable short- and medium-term priorities, intermediate objectives and conditions.

According to the 'Copenhagen criteria', Turkey must attain a strong democratic stability, with protection of human rights and respect for minority rights. Trying to fulfil those criteria, the government created the 'National Programme' in 2001, aimed at bringing Turkish democracy to the level of European democracy by creating a legal foundation for the full protection of individual rights and freedoms, the freedom of thought and expression, the freedom of association and peaceful assembly, and the enlargement of the space of civil society. More recent reforms have concerned abolishment of the death penalty, elimination of the legal restrictions to the rights of different ethnic communities both to education and to broadcasting in their own mother tongues. Moreover, non-Muslim religious foundations have been granted the right to acquire property (Keyman and Içduygu 2003).

In this context, the democratisation process of Turkey has a global aspect which, along with the increasing globalisation of markets and the growth of global communication, has resulted in the disappearance of the differences that used to distinguish government from opposition policies. In other words, one could not easily understand

> the relationship of the state to its civil society with no or minimal recognition of the existence of other states and that of globalisation which gives rise to issues such as environmental risks, multiculturalism, and poverty. Those issues cannot be solved by political parties alone. In Turkey, as much as in other countries, actors such as civil society organisations should act as 'actors of politics' (Keyman and Içduygu 2003: 225).

Today, many CSOs are often funded directly by the EU in order to contribute to the country's effort to meet the EU accession criteria (Beşpınar-Ekici and Gökalp 2006).

This brings into discussion another aspect of the state–civil society relationship. The republican model of citizenship as well as ethnic and religious fundamentalisms inherent in some CSOs can be seen as evidence that link CSOs with certain sections of state power. Keyman and Içduygu (2003) argue that unlike other countries, civil society organisations in Turkey are not just issue-specific organisations; they are rather involved with big societal issues and visions, such as Kemalism, a modern Turkey, the protection of contemporary civilised life, the secular, democratic Turkey or Islamic order, Islamic life and other issues prominent in state agendas. As some scholars claim, civil society in Turkey has functioned as a legitimator of the state's existence.[7] Quoting Navaro-Yashin (2002: 134), 'recently in Turkey, the idea of the state has been significantly sustained through the everyday practices of people outside the centres of official power'. This means that the state has been reproduced even in aspects of life which, at first glance appear to not be related to state functions. As Navaro-Yashin (2002) puts it, the binary analysis of state/civil society is a discourse of power that forms the basis of new ideologies of power in Turkey in their struggle with one another for influence over the state. So enmeshed are discourses of civil society and state on the ethnographic ground that the analytical distinction is obsolete (ibid.: 136).

An incident which I witnessed during my fieldwork illustrates how this articulation becomes conspicuous. During the first months of my research I attended one of the meetings organised by the Bosphorus Arnavutköy Association (*Boğaziçi Arnavutköylüler Derneği*), which also includes participants of the ASG. After the meeting there was a slide projection of pictures taken from Arnavutköy by two amateur photographers. One of those pictures centred on an open window on the opposite wall of which was hanging a picture of Atatürk. The instant that slide appeared virtually everyone in the room began to spontaneously applaud. 'He is our everything,' commented one of my informants later when I was discussing the incident with her. A civil-society organisation was applauding and thus employing in its practice, a figure identified with secular statism more than any other individual figure in the history of the modern Turkish nation. This episode is one of many examples which demonstrates the power that the state in Turkey exercises on individuals in a manner which could be described as 'embodied'.

7 Kubicek (2005) claims that some CSOs have even contributed to the 1980 *coup d'etat*. Furthermore, CSOs have been acting as aid to the state, after the devastating impact of the neo-liberal politics followed since the 1980s which resulted in the shrinking of the welfare state and the threat to the well-being of urban masses in particular (Beşpınar-Ekici and Gökalp 2006).

This evidence compels us to examine civil society not only in relation to or in opposition to the state but also as a substitute for it (Jalali 2002; Beşpınar-Ekici and Gökalp 2006). For example, 'Turkish environmental politics have generally resembled environmental politics in Europe and North America. In many cases, Turkey has turned to European and American governments and nongovernmental organisations such as Greenpeace, for assistance with major environmental improvement projects' (Ignatow 2007: 55). In the course of my fieldwork, I visited many places in Turkey where development and preservation projects funded by the EU were in progress; for example, one location was involved a project on sustainable agriculture. An NGO had been assigned to educate local farmers regarding sustainable agricultural practices, so that the natural resources of the area, which was declared 'a natural protected area', could be preserved from intensive agriculture. The Turkish co-ordinator of the project told me: 'I came here to implement this project which consists of three parts. The first part is the education of the local people about the preservation of the area. I don't care if they agree or not. They are obligated to listen to me.'

As an evaluator for the World Bank stated in 1985 in a staff working paper, private development organisations were superior to other aid agencies in implementing rural development in Latin America. He also suggested that these NGOs were an important and underutilised resource that should be mobilised to supplant the action of the state (Carroll 1992: 1). In other words, the evaluator implied that non-governmental organisations can easily work as state agencies by implementing development policies, rendering civil society an arena used by the state for promoting its policies and, as suggested by the statement of my informant, sometimes this is done in a more authoritative way than methods used by state authorities.

Secularism and the Army

Examples in the Turkish context, such as religion indicate an even less conspicuous articulation between the state and civil society. Turkey's political history, following its establishment as a nation, shows that in the midst of the continuous swinging between democracy and *coups d'état*, economic liberalism and statism and lately between secularism and Islamism, the army seems to have been omnipresent, decisive and omnipotent. It has been and continues to act as guarantor of democracy against any threat which nowadays seems to be Islam. The secular character of the new Turkish state was supported by intellectuals in the early years of the Turkish Republic. Young Turks perceived religion as a private matter and did not see it as part of Western administration models (Karabelias 2004: 32). Later on, reforms regarding religious prohibitions and the establishment of control over religious groups opened the door for a new statist ideology in which religion had no place.

Today, according to the dominant version of modernity, tradition is seen as a hindrance to development and any collectivity which does not share this idea has always been considered an enemy. Turkish citizens should be held away from traditional values and follow the secular example of the Turkish state. The latter, has always been suspicious of any form of religiosity and any religion-based activity is considered as potential threat to the secular character of the Republic (Özdemir 2003).

Based on the above, one might say that secularism is a concept with very strong connotations within Turkish society. Even though religion in Turkey is left completely outside the State, religious issues are highly politicised even more than in countries without state–religion separation. As mentioned earlier, secularism is one of the principles of Kemalism and the Turkish state is a secular oriented state. Separation between church and state exists since the first steps of the Republic. Today, in Turkey, religious attire in public buildings, such as universities, is not allowed.[8] Even though not prohibited, religious weddings are not acknowledged by the state and only civil weddings are registered. In the course of my fieldwork, I have

8 One of the most publicised cases regarding religious attire was the case of the wife of the current President of the Turkish Republic, Abdullah Gül. When her application for admission to the University of Ankara was denied due to religious attire, Hayrünissa Gül appealed the decision at the European Court of Human Rights. The case was highly publicised in Turkey until she decided to withdraw her appeal due to her husband's position as Minister of Foreign Affairs, who could not be both plaintiff and defendant.

encountered individuals who considered that having a religious wedding was sign of backwardness and in their opinion, those ceremonies were performed mainly by 'backward' Kurdish people.

As mentioned in Chapter 3, many of my informants were passionately in favour of the secular character of the Turkish state and they felt safe as long as the army had the power to ensure the preservation of this character. 'The army is the best institution,' claimed one of my informants while we were having tea one day, and this seemed to be the opinion of 88 per cent of the sample of a poll published in August 2003 (Mango 2004). Throughout contemporary Turkish history, the army is presented as the defender of the values of Atatürkism. Whenever one or more of these values were threatened seriously, according to army officials, by the Democratic Party (DP) of Adnan Menderes in the 1960s or by leftists in 1971 and in 1980 and finally by Islamists in 1997, the army was there to intervene and bring things back to order. Most of the coups were supported by the intellectual elite, with the exception of the one in 1980, and today the army still enjoys the trust of the people. While I was in Turkey, the Erdoğan government tried to pass a law to abolish the prohibition of religious attire in universities and demonstrations against it were organised. The demonstrators were asking the army to 'intervene at last!'.

Even though the exclusion of religion from public life shifted with the end of the single-party period, when new parties tried to benefit from the votes of religious public that represented a big percentage of the electoral body (Karabelias 2004), those parties never drew away from the secular principles of Kemalism. Parties which were formed after the coup of 1980 held similar attitudes even the Welfare Party, which was the first Islamist party in the office, declared that the government would base its politics on the principles set by Atatürk. Similar statements have been made by officials of the present ruling party (AKP), which is a more modern version of the Welfare Party. Moreover, the policies followed by secularist governments were also followed by political Islam. Although ostracised from political life for some time, political Islam, managed to return in a new, modern or, as the current prime minister put it, 'Islamist–Democratic' face, and to win since the mid-1990s not only the votes but also the trust of the majority of the electoral body of Turkish citizens, in contrast to the elitist CHP.

According to some people, the fears of a growing radical Islamic presence in Turkey's politics are unfounded and the warnings are mostly sounded by the country's old elite, which is trying to label Reccep Tayyip Erdoğan, Chairman of the Justice and Development Party (AKP) and present Prime Minister of Turkey, as a religious fundamentalist who is exploiting his impoverished and Islamic origins (Özdemir 2003: 14).

However, some – very few – of my informants have told me that as long as the government follows the basic principles of Kemalism and maintains a European perspective for the country, it is not dangerous. As a thirty-four-year-old journalist and active ASG participant put it: 'Erdoğan proved to be much better than everyone expected him to be. I think that what this government is doing is good and I don't care if they believe in God.'

In other words, what my informant implied was that even if AKP has been labelled as an Islamist party, generally it follows the doctrines of Kemalism. However, this view demonstrates how much 'public life'[9] – both people and the state – is infiltrated by the principles of Atatürkism. An Islamist party identified with a civil society quality – religion – and the people, who by definition are considered civil actors, proclaim the same beliefs; the statist secular beliefs, which penetrate many aspects of Turkish society and which a state institution – the army – is charged with protecting. In this sense, it is rather difficult – if not impossible – to consider state and civil society in Turkey as two distinct realms and examine them as such. It is only when we examine them within the same realm and focus on their interaction rather than their separation, that a clear picture of non-state organisations and social movements will emerge.

9 In this book I adopt the notion of 'public life' as defined by Navaro-Yashin (2002: 2) according to whom the 'public' refers both to people and the state and it produces and recasts the political enabling an analysis on both people and the state as the same domain.

Conclusions: 'We don't want to see a third bridge in Arnavutköy or in any other district'[10]

All groups in society become compelled to pursue their interests 'within the domain organised by the state' through political struggles focused on legal categories (Gledhill 1994: 15). Appadurai, for example, while discussing the relationship among Yanomami populations in the rain forests of Brazil and Venezuela, claims that 'even though Yanomami groups are in a position to reproduce their own neighbourhoods (movements), they are increasingly prisoners in the context-producing activities of the nation-state, which makes their own efforts to produce locality seem feeble, even doomed' (1996: 186). Similarly, the ASG does not have the power to establish a discourse. Its environmental discourse is essentially a reproduction of the dominant global environmental discourse which is primarily the discourse used by the state. The difference is only that this reproduction is not doomed for the ASG (as Appadurai claims it is for the Yanomami), because ASG participants have managed to adopt the prerogatives of hegemonic environmentalism and present them as their own prerogatives. Moreover, their claims seem even fairer than the claims of the Ministry of Public Works and Settlement since they are not interest-related – in terms of economic benefits –but linked to environmental and social justice.

With the above discussion as a theoretical context, this book has examined the ASG as a response to or a result of policies. Following this, I reconsidered the view that social movements are part of the so-called civil society and in opposition to the state and its politics. In the development of my argument, I challenged the application of the concept of civil society, which cannot be seen as a separate realm from that of the state. In this sense, social movements – which by definition belong to civil society – cannot be seen as entities unarticulated with the state sphere.

Approaches to civil society in Turkey examine it from different, but not necessarily conflicting, points of view. In fact, what they all have in

10 ASG placard (see Chapter 3).

common is that the point of reference to the development of the non-statist form of action by different groups in society is the aftermath of the 1980 *coup d'état*. In addition, they all use the term civil society, defining according to their opinion what civil society in Turkey is, or could be. They agree, however, that the intervention of the state in civil society is significant and – as Navaro-Yashin (2002) claims – in some cases invisible and thus more influential. The gesture of clapping during a meeting of BAD, which is considered a 'civil society' actor, the minute the picture of Atatürk appeared on the screen, described above, illustrates this intervention in a symbolic level. More than that, as discussed in Chapters 2, 3 and 4 as well as in the present chapter, the statist influence is evident in many aspects of public life in Turkey in general and in ASG in particular. In this sense, people and the state should be analysed as same and not as different domains because 'the political is met not only in sites but also in fleeting and intangible forms' (Navaro-Yashin 2002: 3). These flexible forms place politics in sites that have been conceived as non-political or else as 'civil society' sites.

Within this framework, this ethnographic study has examined a manifested response (ASG) to policy making and the ways in which this manifestation negotiates its relationship with the Turkish state.[11] Moreover, the government's effort to join the EU has created an international environment of political opinion about Turkey. Globalisation, through economic liberalisation and media globalisation have affected Turkish activism. The foundation of international NGOs – such as Greenpeace during the 1980s and 1990s in Turkey – illustrates the above argument. All these organisations including ASG are globally informed organisations – through the changes in information technology, world order and the challenge to national sovereignty (Williams 2003) – and are themselves part of, what Hannerz (1996) calls 'global interconnectedness'. In other words, they are part of 'social economic and demographic processes that not only take place within nations but also transcend them in a way that

11 After all, as Alexander (2002: 2) reminds us political indicates power relations rather than activity constitutive of, and monopolised by, the state.

attention limited to local processes, identities and units of analysis yields incomplete understanding of the local' (Kearney 1995: 547).

Besides, environmentalism in developing countries has been influenced – or dominated (Argyrou 2005) – by ideas derived from developed nations through global civil society and the world media (Ignatow 2008b). These[12] points suggest that an examination of the ASG should not treat it as specifically a Turkish group. The activities of the ASG associate its claims with those of global social movements making it part of them. In other words, ASG is a group with a transnational character, since its concerns are transnationally informed and its environmental rhetoric is at some points similar to that rhetoric of global environmental 'civil' actors.

Nevertheless, the character of the ASG is definitely not determined only by global prerogatives. As in many cases around the world, discussed above and in Turkey (Chapter 4), the ASG is characterised by the association of global demands with local political claims. Ignatow (2008a: 13), in discussing the cases of Turkish Islamist and Alevi environmentalisms, claims that movements affected by globalising forces are forms of hybrid environmental movements. In the case of ASG, ideas of sustainability penetrate the environmental discourse and give it political qualities that go beyond environmental issues. Thus, the idea of sustainable development provides a framework that gives transenvironmental meaning to collective action (Kousis and Eder 2001: 11), and as Behrooz Morvaridi (in Özdemir 2003: 3) reminds us, 'Turkey's environmental problems are social problems, reflecting the interrelationships among resource change, human productive activities, and the accompanying transformations of people's lives.'

Both ASG and all the other similar campaigns such as those discussed in Chapter 4, would agree with this assessment of the interrelationships. Their concordance creates a loose network which has the characteristics of a social movement, parts of which each one of them constitutes. This movement in general and ASG in particular see the environment not as a 'true or false' game. For ASG the environment is a medium used to speak

12 In combination with ASG's arguments, activities and identities that have global aspects.

about other issues as in the festival and the other gatherings that strengthen the ties of the community. In other words as mentioned above, the environment in the case of ASG emerges as a means to speak about power and its control. Saving Arnavutköy is important not just for Arnavutköy *per se* but for the sake of saving the power to 'define the meanings of our micro-world' (Argyrou 2005). Hence, as in the title of this sub-section, the refusal to accept the bridge not only in Arnavutköy but anywhere in Istanbul.

CHAPTER 6

Conclusions: From Environmentalism to Transenvironmentalism

This book has presented a description and analysis of a collective action in contemporary Istanbul, specifically a grass-roots urban protest ranged against a particular development project decided upon by the Turkish government. The analysis began in Chapter 1 by contextualising the protest in contemporary Turkey and Istanbul and moved to a closer ethnographic examination of the protest itself. With a review of the foundation of the Turkish nation-state and focus on the developmental policies followed by the ruling parties since 1923, Chapter 2 examined the effects of these policies in the urban and social environment of Istanbul. The chapter emphasises that from the 1950s Istanbul began a process of transformation into a global city, and in this globalisation, economic and cultural factors are involved in intense interactions and realignments. Neo-liberalisation created new priorities for Istanbul such as the construction of business districts and large highways, including the Bosphorus bridges which facilitate access to them. These developmental choices have created social conflicts such as the Beyoğlu case (Chapter 4) and the Arnavutköy anti-bridge protest – the focus of this work.

At first glance these conflicts might seem to be caused by the local displacements that projects such as the third bridge create. However, a closer ethnographic look suggests that they involve a lot more than NIMBY (Not in My Back Yard) concerns. For ASG participants, the development choices made by past and present governments were mainly informed by big corporate national and international interests and not by ordinary people's needs. This is why projects like the third bridge are not simply undesirable projects in the opinion of the protestors but undertakings which symbolise the state's submission to corporate interests and its disregard for the worldviews and needs of the majority of its citizens.

According to ASG, the views of the citizens are informed by their interest in the preservation of the environment, democracy and social justice – which come into conflict with the technocratic interests of the government. From this perspective, the analysis of the ASG protest has taken into account first of all the environment into which the ASG was born (and by this I mean the *public life* in Turkey, which consists of both state and non-state bodies), and secondly the particular cultural features of ASG. Therefore, in Chapters 3 and 4 I attempted to illustrate the particular character of the ASG as a product of evolving identities, activities taking place and the sense of community created amid the struggle as well as, as a product of the particular character of Turkish *public life* which is informed by all of these concerns.

Chapter 3 focuses on the distinction between the formal and informal face of the ASG; the former refers to the environmentalist face while the latter concerns the identities of its participants and the activities and claims which these identities entail. The environmentalist face of the ASG concerns the formal presentation of the reasons for protesting against the third bridge. These are the potential environmental degradation of Arnavutköy in the case of the construction of the bridge and the demand for sustainable protection of the area as a world cultural heritage site. One of the main formal arguments put forward by the ASG is that the construction of the bridge will create major environmental problems in the area such as air and noise pollution. Supporters also strongly claim that a construction such as the bridge will alter considerably the unique architectural character of Arnavutköy. In this rhetoric, the environment acquires an instrumental role in legitimating the protest, and, similarly, as a legitimate agenda, environmental concerns strengthen the protest. For example, the inclusion of Arnavutköy in the ICOMOS (International Council on Monuments and Sites) 'heritage@risk' list has not only legitimated the protest in the eyes of the international community but has also functioned as a protection shield in domestic decision-making. It could be argued that the ICOMOS initiative has probably affected the initiation of the bridge project. Given the European aspirations of the Turkish state, overlooking such a classification would mean non-compliance with international and European environmental directives, a development which could be a hindrance to the country's accession.

Nevertheless, Environmental protection is not the only formal reason that ASG puts forward. In the course of the anti-bridge campaign environmentalism fuses with demands for democratic participation and protection of human rights. These claims refer to the democratic rights of the residents of Arnavutköy to participate in decision-making concerning their place and consequently this claim relates to respect for human rights. For ASG, decision-making such as that for the construction of the bridge, should take into account the views of the community where the development is planned, and this is how supporters perceive democratic participation. The opposite suggests violation, not only of democracy but of human rights as well.

These last issues also implicate features of the informal aspect of ASG; in other words, the transenvironmental. For ASG participants, doing something positive for Arnavutköy's nature is not their only prerogative. For active participants, a victory over the bridge project would mean that the bridge would be built neither in Arnavutköy nor in any other place (see Chapters 3 and 5). For them, Istanbul does not need another bridge, but a stronger welfare state which would provide high-quality public transportation for all its citizens, and not only – in the name of competitiveness – solutions for those who work downtown. Active participants are politically sensitive individuals who have experienced marginalisation by the Turkish state in the past and, for them, it seems that history is repeating itself. However, now they are more powerful than they had been in the past, a reality they are reminded of by less active participants, who demand that decisions about Arnavutköy should take under consideration that it is not an isolated neighbourhood, but an important place in an important city with important alliances. This second sub-group of ASG consists of people who express the ideals and expectations of the new generation of Turkish activists. They are persons who are aware of their power in a country which is trying to come to terms, both domestically and internationally, with a Western democratic model of governance. Consequently, they call for respect of the ASG as a civil actor and Arnavutköy as a community and as a way of life. Finally, the non-active participants oppose their displacement as much as any other person would oppose being disposed of his or her place of birth, childhood and lifelong residence. For these people, leaving Arnavutköy behind means leaving their lives behind. As one of them told me in an informal conversation: 'The truth is, I was not born here but I grew up here, I got married here, my life is here!'

This book has followed a viewpoint that sees environmentalism as a type of cultural perspective. This favours an analysis of the ASG as a group of people whose claims regarding the environment stem from the ways they understand their world and their position in it. As the sixty-year-old less active participant in Chapter 3 claimed, for him the cultural environment is as important as the physical environment and protecting the environment of Arnavutköy means protecting nature, and people, and the relationships between the two and between the people.

These elements synthesise a profile of the initiative that moves from environmentalism to transenvironmentalism in the sense that the informal aspects of ASG embrace much more than environmental concerns. For this reason, Chapter 3 also focuses on the activities which give ASG its particular cultural and, consequently, transenvironmental features. Activities such as the *panayır*, the grand dances, the tea parties and the discussion panels rendered the protest an instrument of connection between the residents of Arnavutköy. As Ayşe said, she would not have met all those people, some of which became really good friends of hers, if ASG had not been formed. Therefore, the protest itself has become instrumental in the sense that Arnavutköy has become more of a community than it was before. One past supporter of the initiative, a social scientist, claimed that he had stopped being involved with ASG because the danger of the construction was no longer imminent and told me during our interview:

> Question: If the danger is not imminent anymore, why then the residents of Arnavutköy are still alert like before?
>
> Answer: Because the danger is never completely obsolete and because social movements don't cease to exist even if the issue which led to their formation is resolved. Social movements are also about the need of communication and ASG serves that need.

This is another point which supports the plural character of ASG. In fact, as my data suggests the importance that the initiative holds in the neighbourhood seems to stem as much from its community features as from its environmentalist agenda. Moreover, the fusion of the cultural attributes of environmentalism with claims and actions which lack direct reference to the environmental cause (such as claims for democracy, better polity, human rights and social justice) point to another aspect of the transenvironmental character of ASG – the political aspect. Examples from Islamist groups and the Alevi (Chapter 4) support this aspect as well, not only when they function as separate groups, but also because traits of their worldviews can be found in protests like ASG.

Throughout Turkey there are similar protests with similar demands that exceed the issue-specific claims which triggered their formation. Therefore, this book has attempted to examine the ASG as part of social movements which seek to establish or re-establish their relationships with the states in which they are included. Following this, Chapter 4 discusses the broader context to which ASG is a part: the context created by other environmental organisations within the same national context. As Rootes (1999: 34) claims in the case of environmental organisations, 'the national peculiarities of environmental movements clearly testify to the persistent impact of national cultures and political structures and bear the imprint of national policies'. Indeed, the similarities that ASG bears with the campaigns discussed in Chapter 4 suggest that it is not an isolated protest. On the contrary, it is a group like many others existing in contemporary Turkey which share similar concerns because they bear the imprint of Turkish policies. In this sense, it could be claimed that ASG, along with all these groups, forms a transenvironmental social movement that negotiates its very particular relationship with the Turkish state.

Nevertheless, Chapter 4 claims that not all environmental organisations share the same concerns. Therefore, this chapter distinguishes the differences between formal environmental associations and local environmental campaigns like ASG. The differences between these two categories of environmental organisations range from levels of formality to the nature of aspirations and demands. This classification challenges the notion of Turkish civil society as far as its ability to encompass all those diverse organisations is concerned but also the concept of civil society as such. Using examples from both groups of organisations and focusing on the ASG, Chapter 4 concludes that Turkish environmental organisations – both formal associations and informal local campaigns – co-ordinate their actions with the state and function either interactively or against it respectively.

Accordingly, ASG is seen as a response to policies and as a group that negotiates its claims with various levels of state administration. As discussed in Chapter 3, the ASG asks for the support of the local authorities (*muhtarlık*) both of Arnavutköy and Kandilli. In addition, the proclamation against the bridge from candidates for local office during the 2004 local elections suggests that implicitly there is a negotiation between the ASG and the city administration. Finally, the interaction of the protest, in terms of criticism and demanding, with the central administration is evident throughout the entire protest. The environmental claims of the ASG, democratic participation issues, activities of the initiative – from festivals to discussion panels – press releases, informal accounts of the participants and the symbolic gesture to visit the parliament (see Chapter 3), suggest that ASG's relationship with central administration is continuous and a central point of reference for the protest. Moreover, both the interaction with local authorities (*muhtarlıks*, municipalities) and with the central state, suggest the complexity of and the undeniable relationships between the neighbourhood and the state, rendering ASG a group with clear political orientation, part of Turkish *public life* and not a mere 'civil' actor. In this way, like many other grass-roots organisations, ASG becomes a mediator between the state and its citizens.

Moreover, the transenvironmental campaigns discussed in Chapter 4, except from the national level – like ASG does – link their claims with prerogatives of international political bodies. For example, as a local grass-roots mobilisation, the Bergama protest (discussed in Chapter 4) was successful because it connected its demands with concepts such as citizenship, human and environmental rights as well as with global agendas. In this way it gained legitimacy in the national political arena (Beşpınar-Ekici and Gökalp 2006). The Bergama activists aimed to engage citizens in wider issues of political power with the right to engage in legislative and judicial power. They operated through legal channels, recognising accordingly the supremacy and legitimacy of the Turkish Republic. At the same time they did not hesitate to bring the issue to the EU, moving beyond the national borders of their own national constitutional democracy (Öncü and Koçan 2002). In this way, they brought Turkey face to face with the international community and their campaign in the middle of this encounter.

Similarly, the palpable relationship of ASG with the state lies on various levels. Chapter 5 sets the theoretical analysis in a context of the interaction between local campaigns and the state, as well as transnational and supranational bodies. Having argued in the previous chapters that ASG is more than a local protest, this chapter explores the literature on theories of social movements supporting the claims of Chapters 3 and 4 that ASG is part of wider social movements. In addition, the chapter analyses ASG as a result of the effects of globalisation. As discussed in Chapter 3, the campaign in many cases adopted prerogatives stemming from the global environmental discourse. In this sense, the ASG becomes part of the so-called 'global civil society'. ASG's environmental demands: the environmental protection of Bosphorus in general and Arnavutköy in particular, the sustainable development of the neighbourhood, the association of ASG with international environmental organisations, the claims of ASG participants regarding the significance of environmental protection, from all the subgroups identified above suggest that the ASG is, indeed, not just a product of national and local contexts but also a product of globalisation. Chapter 5 argues that cases of contemporary local activism should be looked at as part of wider networks both in terms of space and scope. Thus, the ASG is examined, from this perspective, as a part of social movements which represent the *cultural dimensions of globalisation* and in the context of which *major transforming changes have been occurring* in terms of larger national and transnational politics. However, such a consideration creates yet another level of analysis. One which considers the relationship which the ASG, as a product of Turkish politics, establishes between the Turkish state and transnational bodies such as the EU. Having assimilated global environmental prerogatives, the ASG has become a group which is not only transnational in character but an informal mediator between the Turkish state and the international community as well.

In this sense, ASG's political role is not limited in the national context but also in the transnational arena suggesting, as discussed in Chapter 5, that mainstream notions of civil society are problematic when they refer to either national or transnational contexts. In turn, this assumption leads to the reconsideration of social movements as political rather than civil actors, a re-definition which is pertinent for ASG. This re-consideration of social movements in general and environmental movements in particular as political movements underpins the argument that groups such as the ASG are not merely environmental actions but transenvironmental ones as well.

To sum up, the ASG is a multifaceted group. It is not simply a local protest, not merely an urban movement and definitely not just an environmental campaign. It is a local group that is part of a transenvironmental social movement the qualities of which are shaped by what Falk (1993) calls globalisation-from-below: that is, transnational forces that share common characteristics either when they oppose to a particular development such as the ASG struggle, or when they raise issues of human rights, environment, democracy and other issues such as oppression, poverty, humiliation and collective violence which result from political decisions. In addition, the focus on cultural particularity is another factor which takes the study of environmentalism one step further from a purely environmental debate. Socio-cultural peculiarities are always present as supported by the earlier discussion on Turkish 'civil society' and the identities that evolve in the course of the protest and unify the ASG participants. This approach suggests that the examination of action groups such as the ASG as transenvironmental (groups) should avoid generalisations based on 'universal' meanings. This perspective would help to initiate a discussion of environmental conflicts as societal conflicts, where, the distinct character and the peculiarities of a protest group are viewed as social, political and cultural rather than issue-specific.

References

Akşam. 'İstanbul'da üçüncü köprüye ihtiyaç var mı?' (30 June 2002).
——. 'Üçüncü köprüye Hayır' (7 June 1999).
Alexander, C. *Personal States: Making connections between people and bureaucracy in Turkey* (New York, NY: Oxford University Press, 2002).
Alvarez, S.E., Dagnino, E. and Escobar, A. *Cultures of Politics/Politics of Cultures: Re-Visioning Latin American Social Movements* (Boulder, CO: Westview, 1998).
Anderson, B. *Imagined Communities* (London: Verso, 1983).
Appadurai, A. *Modernity At Large: Cultural Dimensions of Globalization* (Minneapolis, MI: University of Minnesota Press, 1996).
——. 'Grassroots Globalization and the research imagination'. *Public Culture* 12 (1) (2000), 1–19.
Argyrou, V. *The Logic of Environmentalism: Anthropology, Ecology and Postcoloniality* (London: Berghahn Books, 2005).
Arundhati, R. *I aksìa tis zois* (Athens: Psichoyos, 2000).
Arsel, M. 'Risk Society at Europe's Periphery? The Case of the Bergama Resistance in Turkey'. In Gökşen, F. et al. (eds). *Integrating and Articulating Environments: A Challenge for Northern and Southern Europe* (Lisse: Swets and Zeitlinger, 2003).
——. *Transnational Advocacy Networks in the Mediterranean: The case of nuclear energy in Turkey* (n.d.).
Bartu, A. 'Who Owns the Old Quarters? Rewriting Histories in Global Era'. In Keyder, Ç. (ed.). *Istanbul between the Global and the Local* (Lanham, MD: Rowman & Littlefield, 1999).
Bauman, Z. 'The Left as the Counter-Culture of Modernity'. *Telos* 70 (1986), 81–93.
Bender, B. 'Introduction: Landscape – Meaning and Action'. In Bender, B. (ed.). *Landscape: Politics and Perspectives* (Oxford: Berg, 1995).

Bennett, V. 'The Evolution of Urban Popular Movements in Mexico between 1968 and 1988'. In Escobar, A. and Alvarez, S.E. (eds). *The Making of Social Movements in Latin America: Identity, Strategy, and Democracy* (Boulder, CO: Westview, 1992).

Beşpınar-Ekici, U.F. and Gökalp, D. 'Power Brokers, Local Beneficiaries and Empowering the Powerless: A Critical Assessment of the NGOs, Civil Society and Democracy in Turkey'. Paper presented in the Graduate Student Conference, Institute for International and Regional Studies, Princeton University, 7–8 April 2006.

Bliss, S. *Mega Cities: Istanbul – Asian and European City*, accessed at http://science.uniserve.edu.au/school/curric/stage6/geog/ISTANBUL.pdf

Briassoulis, H. 'Sustainable Development – the Formal or Informal Way?' In Kousis, M. and Eder, K. (eds). *Environmental Politics in Southern Europe: Actors, Institutions and Discourses in a Europeanizing Society* (Dordrecht: Kluwer Academic Publishers, 2000).

Brulle, R.J. *Agency, Democracy and Nature: The U.S. Environmental Movement from a Critical Theory Perspective* (Cambridge, MA: MIT Press, 2000).

Bulaç, A. *Din ve Modernizm* (Istanbul: ız Yayıncılık, 1995).

Burdick, J. *Blessed Anastacia: Women, Race, and Popular Christianity in Brazil* (New York, NY: Routledge, 1998).

Burnham, P. 'Whose Forest? Whose Myth? Conceptualisations of Community Forests in Cameroon'. In Abramson, A. and Theodossopouos, D. (eds). *Land, Law and Environment: Mythical Land, Legal Boundaries* (London, Sterling, VA: Pluto Press, 2000).

Canel, E. 'Democratization and the Decline of Urban Social Movements'. In Escobar, A. and Alvarez, S.E. (eds). *The Making of Social Movements in Latin America: Identity, Strategy, and Democracy* (Boulder, CO: Westview, 1992).

Carroll, T. *Intermediary NGOs: The Supporting Link in Grassroots Development* (Sterling, CT: Kumarian Press, 1992).

Castells, M. *The City and the Grassroots: A Cross-cultural Theory of Urban Social Movements* (Berkeley, CA: University of California Press, 1983).

Çelik, Z. *The Remaking of Istanbul: Portrait of an Ottoman City in the Nineteenth Century* (Berkeley, CA: University of California Press, 1993).
Christides, C. *Ta Septemvrianà* (Athens: Centre of Asia Minor Studies, 2000).
Cladwell, L.K. *Between Two Worlds: Science, the environmental movement and policy choice* (Cambridge, New York, NY, Victoria: Cambridge University Press, 1990).
Clark, J. *Democratizing Development: The Role of Voluntary Agencies* (London: Earthscan, 1995).
Cotgrove, S. 'Environmentalism and Utopia'. *Sociological Review* 24 (1976), 23–42.
Cowan, J. *I Politikì tou Sòmatos: Choròs ke kinonikòtita stin Vòria Ellada* (Athens: Aleksandria, 1990).
Cumhuriyet. '3. köprüye panayırlı protesto' (1 February 1999).
——. 'Kentin boyunca kement atılıyor' (5 May 1999).
——. 'Boğaz'a Üçüncü Köprüyü Kimler İstiyor?' (14 February 1999).
Cummings, B. *Dam the rivers, damn the people. Development and resistance in Amazonian Brazil* (London: Earthscan, 1990).
Cunningham, H. 'The Ethnography of Transnational Social Activism: Understanding the Global as Local Practice'. *American Ethnologist* 26 (3) (1999), 583–604.
Dahrendorf, R. *Reflection on the revolution in Europe* (London: Chatto & Windus, 1990).
Danışman, G. and Üstün, İ. 'Yerel Tarihin Gücü: Arnavutköy Semt Girişimi'nin 3. Boğaz Köprüsü Projesi ne Yerl Tarih ile Karşı Koyusu'. Paper presented at the Third International Historical Conference: New Approaches to Historical Writing and Museology: Globalisation and Localisation (Üçüncü Uluslararası Tarih Kongresi. Tarih Yazımı ve Müzecilikte Yeni Yaklasımlar: Küreselleşme ve Yerelleşme). Istanbul, December 2000a.
——. 'Arnavutköy 1900–2000: Urban Transformation at a Multi-ethnic Neighbourhood of Istanbul as Evidenced Through Narratives of its Senior Citizens'. In *Crossroads of History: Experience, Memory,*

Orality. Vol. II, XIth International Oral History Conference, Istanbul 2000b.
——. 'Contentious Politics at a Bosphorus Neighbourhood. Perspectives on Conflict and Solidarity during the 20th Century'. Paper presented at the conference on Contentious Politics and the Economic Opportunity Structure: Mediterranean Perspectives. University of Crete, Rethimno, 17–18 October 2003.
Derek, W. *Earth First! and the Anti-Roads Movement: Radical Environmentalism and Comparative Social Movements* (London, New York, NY: Routledge, 1999).
Doane, M. 'The Resilience of Nationalism in a Global Era: Megaprojects in Mexico's South'. In Nash, J. (ed.). *Social Movements: An Anthropological Reader* (Oxford: Blackwell, 2005).
Drijver, C. 'People's participation in environmental projects'. In Croll, E. and Parkin, D. (eds). *Bush Base Forest Farm: Culture, Environment and Development* (London: Routledge, 1992).
Durrenberger, E.P. 'Global Processes, Local Systems'. *Urban Anthropology* 32 (3–4) (2003), 253–79.
Edelman, M. 'Social Movements: Changing Paradigms and Forms of Politics'. *Annual Review of Anthropology* 30 (2001), 285–317.
Edwards, M. and Hulme, D. 'Conclusion: Too Close to the Powerful, Too Far from the Powerless?' In Edwards, M. and Hulme, D. (eds). *NGOs, States and Donors: Too Close for Comfort?* (Basingstoke: Macmillan, 1997).
Einarsson, N. 'All animals are equal but some are cetaceans: conservation and culture conflict'. In Milton, K. (ed.). *Environmentalism: The View from Anthropology* (London, New York, NY: Routledge, 1993).
Emberlin, J.C. *Eisagogì stin Oikologìa* (Athens: TYPOTHYTO, 1996).
Escobar, A. and Alvarez, S.E. 'Introduction: Theory and Protest in Latina America Today'. In Escobar, A. and Alvarez, S.E. (eds). *The Making of Social Movements in Latin America: Identity, Strategy, and Democracy* (Boulder, CO: Westview, 1992).
Escobar, A. 'Construction Nature: Elements for a post-structural political ecology'. In Peet, R. and Watts, M. (eds). *Liberation Ecologies:*

Environment, development, social movements (London, New York, NY: Routledge, 1996).

Esman, M.R. 'Festivals, Change, and Unity: The Celebration of Ethnic Identity among Louisiana Cajuns'. *Anthropological Quarterly* 55 (4) (1982), 199–210.

Fairhead, J. and Leach, M. *Misreading the African landscape: Society and ecology in a forest-savanna mosaic* (Cambridge: Cambridge University Press, 1996).

Falk, R. 'The making of global citizenship'. In Childs, J.B., Brecher, J. and Cutler, J. (eds). *Global Visions: Beyond the New World Order* (Boston, MA: South End, 1993).

Featherstone, M. *Global Culture: Nationalism, Globalisation and Modernity* (London: Sage, 1990).

Feld, S. and Basso, H. 'Introduction'. In Feld, S. and Basso, H. (eds). *Senses of Place* (Santa Fe, CA: School of American Research Press, 1996).

Findji, M.T. 'From Resistance to Social Movement: The Indigenous Authorities Movement in Colombia'. In Escobar, A. and Alvarez, S.E. (eds). *The Making of Social Movements in Latin America: Identity, Strategy, and Democracy* (Boulder, CO: Westview, 1992).

Flogaiti, E. *Perivallontikì Ekpèdefsi* (Athens: Ellinika Grammata, 1998).

Foucault, M. *Power/Knowledge: Selected Interviews and Other Writings 1972–1977*, ed. G. Colin (Brighton: Harvester Press, 1980).

Friedman, J. and Wolff, G. 'World city formation: An agenda for research and action'. *International Journal of Urban and Regional Research* 6 (1982), 309–44.

Geertz, C. *The Interpretation of Cultures: Selected Essays* (London: Fontana Press, 1993).

Gèfou-Madianoù, D. 'Eisagogí: Apó tin Olistikí Proséggisi stis Merikés Alítheies'. In Gèfou-Madianoù, D. (ed.). *Anthropologikí Theoría ke Ethnografía* (Athens: Ellinikà Gràmmata, 1998).

Gellner, E. *Conditions of Liberty: Civil Society and Its Rivals* (London: Penguin, 1994).

Georgoulas, S. 'The Boy Scouts Association of Greece and Ecological Action as a Conservative Practice'. *Nature, Society and Thought* 19 (4) (2006), 427–45.

Geralch, L.P. 'The Structure of Social Movements: Environmental Activism and its Opponents'. In Freeman, J. and Lanham, V. (eds). *Waves of Protest: Social Movements Since the Sixties* (Lanham, MD: Rowman & Littlefield, 1999).

Gledhill, J. *Power and its Disguises: Anthropological Perspectives on Politics* (London: Pluto Press, 1994).

Göle, N. 'Towards an Autonomization of Politics and Civil Society in Turkey'. In Heper, M. (ed.). *Politics in the Third Turkish Republic* (Boulder, CO: Westview Press, 1994).

Guha, R. and Martinez-Alier, J. *Varieties of Environmentalism: Essays North and South* (London: Earthscan, 1997).

Gulick, J. 'The Essence of Urban Anthropology: Integration of Micro and Macro Research Perspectives'. *Urban Anthropology* 13 (2–3) (1984), 295–306.

Gürbey, S. 'Civil Society and Islam in Turkey'. Paper presented at the Graduate Conference of Columbia University, Spring 2006.

Gürsel, D.E.G. *Çevresizsiniz* (Istanbul: InsanYayınları, 1989).

Hall, A. 'Grassroots Action for Resettlement Planning: Brazil and Beyond'. *World Development* 22 (12) (1994), 1793–1809.

Hall, J. 'In Search of Civil Society'. In Hall, J. (ed.). *Civil Society: Theory, History, Comparison* (Cambridge: Polity Press, 1995).

Hamel, P., Lustiger-Thaler, H. and Mayer, M. 'Introduction: urban social movements – local thematics, global spaces'. In Hamel, P., Lustiger-Thaler, H. and Mayer, M. (eds). *Urban Movements in a Globalizing World* (London, New York, NY: Routledge, 2000).

Hann, C. 'Philosophers' Models on the Carpathian Lowlands'. In Hall, J. (ed.), *Civil Society: Theory, History, Comparison* (Cambridge: Polity Press, 1995).

Hannerz, U. *Exploring the City: Inquiries Toward and Urban Anthropology* (New York, NY and Guildford: Columbia University Press, 1983).

——. *Transnational Connections: Culture, people, places* (London: Routledge, 1996).

Harbers, H. 'Book reviews: *We Have Never Been Modern*'. *Science, Technology & Human Values* 20 (2) (1995), 270–6.

Haynes, J. 'Power, Politics and Environmental Movements in the Third World'. In Rootes, C. (ed.). *Environmental Movements: Local, National and Global* (London: Frank Cass, 1999).
Herzfeld, M. *A Place in History* (Princeton, NJ and Oxford: Princeton University Press, 1991).
——. 'Hellenism and Occidentalism: The Permutations of Performance in Greek Bourgeois Identity'. In Carrier, J. (ed.). *Occidentalism: Images of the West* (Oxford: Oxford University Press, 1995).
Ho Kwok-Leung, D. *Polite Politics: A sociological analysis of an urban protest in Hong Kong* (Aldershot: Ashgate, 2000).
Hürriyet. '3. Köprü zararlı' (4 May 1999).
Ignatow, G. 'From Science to Multiculturalism: Postmodern Trends in Environmental Organizations'. *Global Environmental Politics* 5 (2) (2005a), 88–113.
——. 'Economic Dependency and Environmental Attitudes in Turkey'. *Environmental Politics* 14 (5) (2005b), 648–66.
——. *Transnational Identity Politics and the Environment* (Lanham, MA: Lexington Books, 2007).
——. 'Globalizations and the Transformation of Environmental Activism: Turkey since the 1980s'. *Globalizations* 5 (3) (2008a), 433–47.
——. 'Transnational Environmentalism at Europe's Boundaries: Identity Movements in Lithuania and Turkey'. *Current Sociology* 56 (6) (2008b), 845–64.
Ingold, T. 'Globes and spheres: the topology of environmentalism'. In Milton, K. (ed.). *Environmentalism: The View from Anthropology* (London, New York, NY: Routledge, 1993).
ITÜ ve İstanbul Büyük Şehir Belediyesi. 'İstanbul Ulaşım Ana Planı' (Istanbul Traffic Master Plan) (1998).
Jalali, R. 'Civil Society and the State: Turkey after the Earthquake'. *Disasters* 26 (2) (2002), 120–39.
Jamison, A., Eyerman, R., Cramer, J. and Læssøe, J. *The Making of the New Environmental Consciousness: A Comparative Study of the Environmental Movements in Sweden, Denmark and the Netherlands* (Edinburgh: Edinburgh University Press, 1990).

Kadioğlu, A. 'Civil Society, Islam and Democracy in Turkey: A Study of Three Islamic Non-Governmental Organizations'. *The Muslim World* 95 (2005), 23–41.

Kandiyoti, D. 'Introduction'. In Kandiyoti, D. and Saktanber, A. (eds). *Fragments of Culture: The Everyday of Modern Turkey* (London and New York, NY: I.B. Tauris, 2002).

Karabelias, Y. *Tourkìa: Thriskìa, Koinonia ke Politikì* (Athens: Ellinikà Gràmmata, 2004).

Karafotakis, E. *I Diamòrfosi tis Sìghronis Tourkikìs Oikonomìas* (Athens: Kèntro Politikìs Èrevnas ke Epikoinonìas, 2000).

Kasapoğlu, M.A. and Ecevit, M.C. 'Attitudes and Behaviour Toward the Environment: The Case of Lake Burdur in Turkey'. *Environment and Behaviour* 34 (3) (2002), 363–77.

Kasmir, S. 'Activism and Class Identity: The Saturn Auto Factory Case'. In Nash, J. (ed.). *Social Movements: An Anthropological Reader* (Oxford: Blackwell, 2005).

Kearney, M. 'The Local and the Global: The Anthropology of Globalization and Transnationalism'. *Annual Review of Anthropology* 24 (1995), 547–65.

Kemper, M. 'Trends in urban anthropology research: an analysis of the journal *Urban Anthropology*, 1972–91'. *Urban Anthropology and Studies of Cultural Systems and World Economic Development* 20 (1991b), 373–84.

Kempton, W., Fireston, J., Lilley, J., Rouleau, T. and Whitaker, P. 'The Offshore Wind Power Debate: Views from Cape Cod'. *Coastal Management* 33 (2005), 119–49.

Kenny, S. 'The Hindmarsh Island Bridge: Case-Implications for Native Title'. Paper presented at the AIC Native Title Conference, April 1998.

Kentor, J., Smith, D. and Timberlake, M. 'The World-System's City System: A Research Agenda', accessed at http://www.irows.ucr.edu/conferences/globgis/papers/Smith.htm

Keyder, Ç. 'A Tale of Two Neighbourhoods'. In Keyder, Ç. (ed.). *Istanbul between the Global and the Local* (Lanham, MD: Rowman & Littlefield, 1999).

Keyman, E.F. and Içduygu, A. 'Globalization, Civil Society and Citizenship in Turkey: Actors, Boundaries and Discourses'. *Citizenship Studies* 7 (2) (2003), 219–34.

KHRP. 'The Ilisu Dam: A Human Rights Disaster in the Making' (November 1999).

Kousis, M. 'Greece'. In Rootes, C. (ed.). *Environmental Protest in Western Europe* (Oxford: Oxford University Press, 2004).

Kousis, M. and Eder, K. 'Introduction: EU policy-making, local action, and the emergence of institutions of collective action. A theoretical perspective on Southern Europe'. In Kousis, M. and Eder, K. (eds). *Environmental Politics in Southern Europe: Actors, Institutions and Discourses in a Europeanizing Society* (Dordrecht: Kluwer Academic Publishers, 2001).

Kubicek, P. 'The European Union and Grassroots Democratization in Turkey'. *Turkish Studies* 6 (3) (2005), 361–77.

Kutsche, P. 'World System Theory as an Integrator of the Social Science Curriculum'. *Urban Anthropology* 18 (1) (1989), 129–34.

Latour, B. *We Have Never Been Modern* (New York, NY, London: Harvester Wheatsheaf, 1993).

Lewellen, T.C. *Political Anthropology: An Introduction* (Westport, CT: Praeger Publishers, 2003).

Lober, D.J. 'Why not here? The importance of context, process, and outcome on public attitudes towards the sitting of waste facilities'. *Society and Natural Resources* 9 (4) (1996), 375–94.

Los Angeles Times. 'A Bridge that Turks Refuse to Cross' (1999).

Lowe, S. *Urban Social Movements: The City After Castells* (London: Macmillan, 1986).

Macnaghten, P. and Urry, J. 'Introduction'. In Macnaghten, P. and Urry, J. (eds). *Bodies of Nature* (London: Sage, 2001).

Mango, A. *The Turks Today* (New York, NY: Overlook Press, 2002).

——. *Atatürk: The Biography of the Founder of Modern Turkey* (New York, NY: Overlook Press, 2004).

Marangudakis, M. 'Nature and Power: A Study of the Social Construction of Nature in Eurasia from the Stone Age to the Hellenistic Times'. Unpublished PhD thesis, McGill University, 1999.

Marcus, G. 'Ethnography In/Of the World System: The Emergence of Multi-Sited Ethnography'. *Annual Review of Anthropology* 24 (1995), 95–117.

Mardin, Ş. 'Civil Society and Islam'. In Hall, J. (ed.). *Civil Society: Theory, History, Comparison* (Cambridge: Polity Press, 1995).

Martell, L. *Ecology and Society: an introduction* (Cambridge: Polity Press, 1995).

Martin, P.L. *The Globalization of Contentious Politics: The Amazonian Indigenous Rights Movements* (Oxford: Routledge, 2003).

Marx, K. and Engels, F. *To Kommunistikó Manifésto* (Athens: Courier Ekdotikí, 1998 [1913]).

Mayer, M. 'The fragmentation of social movements and social justice: beyond the traditional forms of localism'. In Hamel, P., Lustiger-Thaler, H. and Mayer, M. (eds). *Urban Movements in a Globalizing World* (London, New York, NY: Routledge, 2000).

Masashiro, Y., Toshimitsu, M. and Mitsubiro, K. 'Repair of the "Golden Horn Bridge" in Turkey'. *IHI Engineering Review* 36 (2) (2001), 219–27.

Mavrìdis, S. 'Epichìrisi, Perivàllon ke Kinitopìisi: To paràdigma tis Metallourgìas Chrisoù TVX Gold ston Strimonikò Kòlpo'. Unpublished BA thesis, Department of Environmental Studies, University of the Aegean, Mytilene, 2003.

McCarthy, J. 'Constraints and opportunities in adopting, adapting, and inventing'. In McAdam, D., McCarthy, J.D. and Zald, N.M. (eds). *Comparative perspectives on social movements: Political opportunities, mobilizing structures, and cultural framings* (Cambridge: Cambridge University Press, 1996).

McCormick, J. *The Global Environmental Movement* (Chichester: Wiley, 1995).

McCormick, S. 'The Brazilian Anti-dam Movement'. *Organization & Environment* 19 (3) (2006), 321–46.

McKay, G. *Senseless Acts of Beauty: Cultures of Resistance* (London: Verso, 1996).

Melucci, A. *Nomads of the Present: Social Movements and Individual Needs in Contemporary Society* (Philadelphi, PA: Temple University Press, 1989).

Milliyet. 'Çevre Koalisyonu' (1 February 1989).
——. 'Boğaz'ımızı sıkmayın' (11 June 2001).
——. 'Yiğit: 3. köprüyü yapacağız' (12 June 2001).
Milton, K. 'Introduction: Environmentalism and anthropology'. In Milton, K. (ed.). *Environmentalism: The View from Anthropology* (London, New York, NY: Routledge, 1993).
——. *Environmentalism and Cultural Theory: exploring the role of anthropology in environmental discourse* (London, New York, NY: Routledge, 1996).
Moore, D. 'Marxism, Culture, and Political Ecology: Environmental Struggles in Zimbabwe's Eastern Highlands'. In Peet, R and Watts, M. (eds). *Liberation Ecologies: Environment, development, social movements* (London, New York, NY: Routledge, 1996).
Muammer, T. *Public Environmental Attitudes in Turkey*, accessed at http://www.inter-disciplinary.net/ptb/ejgc/ejgc3/tuna%20paper.pdf
Nash, J. 'Introduction: Social Movements and Global Processes'. In Nash, J (ed.). *Social Movements: An Anthropological Reader* (Oxford: Blackwell, 2005).
Navaro-Yashin, Y. *Faces of the State: Secularism and Public Life in Turkey* (Princeton, NJ and Oxford: Princeton University Press, 2002a).
New York Times. 'A New Bridge? Not in My Backyard!' (6 February 1999).
Olson, M. *The Logic of Collective Action: Public Goods and the Theory of Groups* (Cambridge, MA and London: Harvard University Press, 1965).
Öncü, A. 'The myth of the "ideal home" travels across cultural borders to Istanbul'. In Öncü, A. and Weyland, P. (eds). *Space, Culture and Power: New Identities in Globalising Cities* (London and Trenton, NJ: Zed Books, 1997).
Öncü, A. and Koçan, G. 'Political Identity and Citizenship Politics. Nationalism and Cosmopolitanism in the Bergama Movement in Turkey'. *New Perspectives on Turkey* 26 (2001), 29–57.
——. *Democratic Citizenship Movements in the Context of Multi-layered Governance: The Case of the Bergama Movement* (Istanbul: Sabancı University and Bilkent University, 2002).

Özdemir, İ. 'The Development of Environmental Consciousness in Modern Turkey'. In Foltz, R. (ed.). *Environmentalism in the Muslim World* (Boston: MIT Press, 2003).

Pamuk, O. *Istanbul: Pòli ke anamǹisis* (Athens: Okeanìda, 2005).

Peet, R. and Watts, M. 'Liberation Ecology: Development, sustainability, and environment in an age of market triumphalism'. In Peet, R. and Watts, M. (eds). *Liberation Ecologies: Environment, development, social movements* (London, New York, NY: Routledge, 1996).

Pickvance, K. 'The diversity of Eastern European social movements: urban movements, new social movements and nationalist movements'. In Hamel, P., Lustiger-Thaler, H. and Mayer, M. (eds). *Urban Movements in a Globalizing World* (London, New York, NY: Routledge, 2000).

Pilar-García, M. 'The Venezuelan Ecology Movement: Symbolic Effectiveness, Social Practices, and Political Strategies'. In Escobar, A. and Alvarez, S.E. (eds). *The Making of Social Movements in Latin America: Identity, Strategy, and Democracy* (Boulder, CO: Westview, 1992).

Piven, F.F. and Cloward, R.A. *Poor People's Movements: Where they Succeed, How they Fail* (New York, NY: Pantheon, 1977).

Psimitis, M. *Isagogí sta síghrona kinonikà kinímata* (Athens: Atrapòs, 2006).

Rangan, H. 'From Chipko to Uttarangal: Development, environment, and social protest in the Garhwal Himalayas, India'. In Peet, R. and Watts, M. (eds). *Liberation Ecologies: Environment, development, social movements* (London, New York, NY: Routledge, 1996).

Raptis, N. *Perivallontikì Ekpèdefsi ke Aghoyì. To Theoritikò Plèsio ton Epiloghòn* (Athens: TYPOTHYTO, 2000).

Robins, K. 'Interrupting Identities: Turkey/Europe'. In Hall, S. and du Gay, P. (eds). *Questions of Cultural Identity* (London: Sage, 1996).

Robinson, M. 'Privatising the Voluntary sector: NGOs as public Service Contractors'. In Edwards, M. and Hulmes, D. (eds). *NGOs, States and Donors: Too Close for Comfort?* (Basingstoke: Macmillan, 1997).

Rollwagen, J.R. 'The City as Context: A Symposium'. *Urban Anthropology* 4 (1) (1975), 1–25.

Rona, B. *Turkish in Three Months* (London: Hugo Press, 1999).

Rootes, C. 'Acting Globally, Thinking Locally? Prospects for a Global Environmental Movement'. In Rootes, C. (ed.). *Environmental Movements: Local, National and Global* (London: Frank Cass, 1999).

Rumelili, B. 'Civil Society and the Europeanization of Greek–Turkish Cooperation'. *South European Society & Politics* 10 (1) (2005), 45–56.

Sabah. '"3" üncü Boğaz geçisi yine uyutuldu mu?' (20 September 1997).

——. 'Başkan bile yanlış biliyor' (26 January 1999).

——. 'Sizi düşüncesiz şehir plancıları sizi!' (18 June 2002).

Sachs, A. 'What do human rights have to do with environmental protection? Everything'. Paper prepared for the Local Libraries: Global Awareness Project, 1997.

Sakellari, M. 'Aksiològisi tis Perivallontikìs Eyaisthitopìisis dimosiogràfon sta topikà M.M.E. tis Mytilìnis'. Unpublished BSc dissertation, Department of Environmental Studies, University of the Aegean, 2001.

Salamon, L.M., Sokolowski, S.W. and List, R. *Global Civil Society: An Overview* (Baltimore, MD: Johns Hopkins University Press, 2003).

Sanjek, R. 'Urban Anthropology'. In Barnard, A. and Spencer, J. (eds). *Encyclopedia of Social and Cultural Anthropology* (London: Routledge, 2004).

Sassen, S. *The Global City* (Princeton, NJ: Princeton University Press, 2001).

Schönwälder, G. *Linking Civil Society and the State: Urban Popular Movements, the left, the local government in Peru, 1980–1992* (University Park, PA: Pennsylvania State University Press, 2002).

Schuurman, F.J. 'Urban social movements: between regressive utopia and socialist panacea'. In Schuurman, F.J. and Van Naerssen, T. (eds). *Urban Social Movements in the Third World* (London, New York, NY: Routledge, 1989).

Shankland, D. *The Alevis in Turkey: The Emergence of a Secular Islamic Tradition* (London, New York, NY: Routledge, 2003).

Simmonet, D. *Tì ìne Oikologhìa?* (Athens: Nèa Sìnora, 1985).

Şimşek, S. 'New Social Movements in Turkey Since 1980'. *Turkish Studies* 5 (2) (2004), 111–39.

Spyridaki, H. 'Melèti Tou Kathestòtos Kataskeuìs Megàlon Fragmàton sti Vrazilìa ke stin Indìa. Vasikà Simìa Sygkritikìs Aksiològisis ton Kinimàton Enàntia sta Fràgmata stis Chòres autès'. Unpublished BSc thesis, University of the Aegean, 2007.

Star. 'Boğaziçi'nde çevre şenliği yapıldı' (7 June,1999).

Stephen, L. 'Gender, Citizenship, and the Politics of Identity'. In Nash, J. (ed.). *Social Movements: An Anthropological Reader* (Oxford: Blackwell, 2005).

Stern, P.C., Dietz, T., Abel, T., Guagnano, G.A. and Kalof, L. 'A Value-Belief-Norm Theory of Support for Social Movements: The Case of Environmentalism'. *Human Ecology Review* 6 (2) (1999), 81–97.

Stirling, P. 'Introduction: Growth and Changes. Speed, Scale, Complexity'. In Stirling, P. (ed.). *Culture and Economy: Changes in Turkish Villages* (Huntingdon: Eothen Press, 1993).

Stott, P. 'Biogeography and Ecology in Crisis: The Urgent Need for a New Metalanguage'. *Journal of Biogeography* 25 (1) (1998), 1–2.

Tarrow, S. *Power in Movement: Social Movements and Contentious Politics* (Cambridge: Cambridge University Press, 1998).

Theodossopoulos, D. 'Turtles, Farmers, and "Ecologists": The Cultural Reason behind a Community's Resistance to Environmental Conservation', *Journal of Mediterranean Studies* 7 (2) (1997), 250–67.

——. 'The Land People Work and the Land the Ecologists Want: Indigenous Land Valorisation in a Greek Island Community Threatened by Conservation Law'. In Abramson, A. and Theodossopouos, D. (eds). *Land, Law and Environment: Mythical Land, Legal Boundaries* (London, Sterling, VA: Pluto Press, 2000).

——. 'Environmental conservation and indigenous culture in a Greek island community: the dispute over the sea turtles'. In Chatty, D. and Colchester, M. (eds). *Conservation and indigenous mobile peoples: displacement, forced settlement and sustainable development* (Oxford: Berghahn, 2002).

Tilley, C. *A Phenomenology of Landscape: Places, Paths and Monuments* (Oxford: Berg, 1994).

Toprak, B. 'Islamist Intellectuals: Revolt against Technology and Industry'. In Heper, M., Öncü, A. and Kramer, H. (eds). *Turkey and the West: Changing Political and Cultural Identities* (London: I.B. Tauris, 1993).

Toth, J. 'Local Islam Gone Global: The Roots of Religious Militancy in Egypt and its Transnational Transformation'. In Nash, J. (ed.). *Social Movements: An Anthropological Reader* (Oxford: Blackwell, 2005).

Tuncer, G., Ertepinar, H., Tekkaya, C. and Sungur, S. 'Environmental attitudes of young people in Turkey: effects of school type and gender'. *Environmental Education Research* 11 (2) (2005), 215–33.

Turkish constitution, 2001.

Turkish Daily News. 'Arnavutköy residents don't want the third bridge' (14 June 2002).

United Nations. *Report of the World Commission on Environment and Development* (1987).

van der Heijden, H.A. 'Environmental Movements, Ecological Modernisation and Political Opportunity Structures'. In Rootes, C. (ed.). *Environmental Movements: Local, National and Global* (London: Frank Cass, 1999).

van der Linden, J. 'The limits of territorial social movements: the case of housing in Karachi.' In Schuurman, F. and van Naerssen, T. (eds). *Urban Social Movements in the Third World* (London, New York, NY: Routledge, 1989).

van Naerssen, T. 'Continuity and change in the urban poor movement of Manila, the Philippines'. In Schuurman, F. and van Naerssen, T. (eds). *Urban Social Movements in the Third World* (London, New York, NY: Routledge, 1989).

Voulvouli, A. 'Environmentalism as Social Critique: The case of ASG'. Paper presented at the International Conference on Nature, Science and Social Movements, Department of Sociology, University of the Aegean, 25–8 June 2004a.

——. 'From "NIMBYism" to "Transenvironmentalism": The struggle against the Third Bosphorus Bridge'. Paper presented at the EASA

8th Bi-annual Conference: Face to face: Connecting Distance and Proximity, University of Vienna, 8–12 September 2004b.
Wallerstein, I. *The Modern World System: Capitalist Agriculture and the Origins of the European World Economy in the Sixteenth Century* (New York, NY: Academic Press, 1974).
Wapner, P.K. *Environmental Activism and World Civic Politics* (New York, NY: State University of New York Press, 1996).
Warren, K.B. and Jackson, J.E. 'Introduction: Studying Indigenous Activism in Latin America'. In Warren, K.B. and Jackson, J.E. (eds). *Indigenous Movements, Self Representation, and the State in Latin America* (Austin, TX: University of Texas Press, 2002).
Washington Post. 'Historic Homes Torn By Times in Istanbul' (2001).
White, J.B. 'The Islamist Paradox'. In Kandiyoti, D. and Sanktaber, A. (eds). *Fragments of Culture: The Everyday of Modern Turkey* (New Brunswick, NJ: Rutgers University Press, 2002).
Williams, M. 'Social Movements and Global Politics'. In Kofman, E. and Youngs, G. (eds). *Globalisation: Theory and Practice* (London: Biddles Ltd, 2003).
Williams, M. and Ford, L. 'The World Trade Organisation, Social Movements and Global Environmental Management'. In Rootes, C. (ed.). *Environmental Movements: Local, National and Global* (London: Frank Cass, 1999).
Wolsink, M. 'Entanglements of interests and motives: assumptions behind the NIMBY-theory on facility siting'. *Urban Studies* 6 (1994), 851–66.
Yalçın, A. 'American Impact on Turkish Social Life (1945–1965)'. *Journal of American Studies of Turkey* 15 (2002), 41–54.
Yalcintan, M.C. and Erbas, A.E. 'Impacts of "Gecekondu" on the Electoral Geography of Istanbul'. *International Labor and Working-Class History* 64 (2003), 91–111.
Yearly, S. 'Standing in for nature: the practicalities of environmental organizations' use of science'. In Milton, K. (ed.). *Environmentalism: The View from Anthropology* (London, New York, NY: Routledge, 1993).
Yeniyüzyıl. 'Köprü möprü is-te-mi-yo-ruz' (15 May 1999).
——. 'Üçüncü köprüye hayır' (22 May 1999).

Web sites

http://adayinlife.typepad.com
http://www.allaboutturkey.com
http://www.arnavutkoy.org
http://www.arnavutkoysemtgirisimi.net
http://www.cevre.org
http://en.wikipedia.org
http://www.evonymos.org/english/medit_sos.html
http://loc.gov.exhibits/marshal
http://www.lse.ac.uk/collections/CCS/introduction.htm
http://www.marmaray.com
http://www.megarevma.net
http://www.panda.org
http://www.politicalquotes.org
http://www.tema.org.tr
http://www.un.org/Conferences/habitat
http://unfccc.int/resource/docs/convkp/kpeng.pdf
http://www.vapurumuvermiyorum.org
http://en.wikipedia.org/wiki/An_Inconvenient_Truth

Index

Alvarez, Sonia 12
anthropology 8, 10, 31, 44, 134, 135
 political 13
 social 1, 8
 of social movements 141
 urban 13, 44
anthropological 2, 10, 45, 52, 66, 120, 134, 135, 139, 141
 study 1, 12
anthropologically 51
Appadurai, Arjun 2, 121, 134, 135, 137, 139, 142, 145, 146, 158
Argyrou, Vassos 58, 119, 128, 160, 161
Arnavutköy 2, 3, 4, 5, 6, 7, 21, 22, 23, 24, 25, 26, 27, 32, 41, 48, 49, 50, 53, 56, 57, 58, 59, 60, 61, 62, 63, 64, 65, 66, 71, 74, 76, 79, 80, 82, 85, 86, 87, 88, 89, 90, 91, 92, 93, 94, 100, 116, 126, 130, 145, 153, 158, 161, 163, 164, 165, 166, 168, 169
Arnavutköy District Initiative (*Arnavutköy Semt Girişimi* – ASG) 2, 4, 5, 6, 7, 11, 13, 15, 17, 25, 26, 27, 28, 29, 32, 36, 43, 47, 48, 49, 50, 51, 53, 55, 56, 58–95, 97–101, 103, 105, 114–17, 119–21, 124, 125, 130, 131, 133, 137, 141, 142, 145, 153, 157–61, 163–70; *see also* Bosphorus Association of Arnavutköy
Arsel, Murat 1, 81, 100, 102, 108, 109, 112, 120, 128, 134
Atatürk, Mustafa Kemal 16, 32–5, 39, 60, 67, 84, 97, 98, 106, 153, 156, 159
Bartu, Ayfer 22, 81, 112
Bauman, Zygmunt 80
Beşpınar-Ekici, Umut F. and Gökalp, Deniz 1, 17–19, 36, 77, 79, 109, 116, 145, 149–54, 168
Bosphorus Association of Arnavutköy 77, 153
Bosphorus Strait 2, 3, 23, 24, 26, 37, 38, 40, 42, 48, 49, 50, 56, 59, 63, 67, 68, 69, 74, 75, 79, 80, 82, 94, 116, 169
 bridge 5, 16, 24, 32, 41, 50, 52, 53, 73, 163
 crossing 16, 25, 41, 43, 73
Bosphorus, University of 73
Bridge, Third 3, 6, 7, 25, 26, 27, 31, 36, 45, 49, 50, 53, 56, 63, 65, 66, 67, 68, 69, 71, 73, 74, 75, 76, 77, 79, 80, 82, 85, 90, 91, 94, 95, 116, 119, 137, 158, 163, 164

Castells, Manuel 15, 27, 44–6, 69, 71, 72
civil society 12, 13, 17, 18, 19, 19, 86, 112, 119, 121, 132, 141, 145–58, 168, 170
 global 79, 106, 143, 144, 169
 organisation 13, 18
 in Turkey 17, 18, 19, 86, 115, 144, 149–61, 167, 170
city
 as a concept 4, 44, 45–7, 71, 72
 fieldwork in 4
 global 5, 25, 31, 32, 37, 47, 49, 76, 163
 see also Istanbul

class
identity 142
middle 38, 76, 79, 82
social 34, 47, 94, 123
struggle 12, 45, 131, 132
working 13, 45, 79
collective action 1, 2, 11, 12, 13, 26, 46, 72, 76, 82, 93, 120, 129, 130, 131, 132, 133, 135, 145, 160, 163
collective actors 138

Danışman, Günhan and Üstün, İsmail 65, 67, 94, 115
Della Porta, Donatella 132
democracy 1, 20, 34, 35, 39, 62, 63, 72, 86, 87, 93, 94, 101, 111, 112, 114, 117, 123, 128, 130, 141, 149, 151, 152, 154, 155, 164, 165, 167, 169, 170
Diani, Mario 132
Durrenberger, Paul 13, 31, 51, 137

Edelman, Marc 10–12, 14, 84, 121, 124, 131–3, 135, 137, 141, 142
Engels, Friedrich 131
environmental
local campaigns 98, 114, 115
local protests 108, 128
local struggles 108
environmentalism 14, 15, 20, 21, 26, 29, 55, 60, 61, 91, 92, 99, 101–3, 106, 107, 119, 121–4, 126, 128, 129, 141, 144, 158, 160, 163, 165–7, 169, 170
Escobar, Arturo 134, 135
Escobar, Arturo and Alvarez, Sonia 12, 134
ethnography 5, 6, 10, 12, 28, 116
multisited 5, 7
ethnographic
accounts 120
approaches 141
data 119
diary 6
examination 163
eye 29, 142
ground 153
look 163
material 29, 98
research 6, 142
space 6
study 12, 159
see also fieldwork, ethnographic

fieldwork 3–6, 8, 10, 14, 44, 52, 61, 68, 77, 82, 84, 85, 88, 153–5
ethnographic 2
Foucault, Michel 147

Geertz, Clifford 66
Gefou-Madianou, Dimitra 10
Gellner, Ernest 146, 149
Georgoulas, Stratos 126
Gledhill, John 13, 147, 158
grass-roots 1, 12, 14, 21, 45, 71, 116, 138, 142, 163, 168; *see also* movement, grass-roots

Hall, Anthony 134
Hall, John A. 147
Hall, Stuart 32
Hamel, Pierre 32, 45–7, 69, 72
Hannerz, Ulf 11, 32, 44, 47, 138, 140, 159
Herzfeld, Michael 80
Hobsbawm, Eric 31

identity 72, 76, 78, 135, 139
of ASG 55, 56, 79, 85
of the ethnographer 8
Islamist 149
national 13
politics 136, 139, 142, 144, 149, 150
secularist 78, 82, 84
transnational 87
see also class identity

Igantow, Gabriel 17, 20, 46, 98, 100, 101, 105–8, 113, 114, 128, 138, 142, 149, 154, 160
Istanbul 1, 2, 4, 5, 6, 7, 16, 17, 22, 23, 24, 25, 29, 31, 32, 33, 34, 35, 36, 37, 38, 39, 40, 41, 42, 43, 45, 46, 47, 48, 49, 50, 51, 52, 53, 56, 57, 58, 59, 63, 69, 70, 71, 72, 73, 75, 83, 87, 90, 91, 94, 106, 112, 125, 161, 163, 165

Kearney, Michael 160
Kemper, Robert V. 4, 5, 13
Keyder, Çaglar 16, 23, 24, 40, 83
Keyman, E. Fuat and Içduygu, Ahmet 17, 18, 98, 145, 149, 150, 152
Kousis, Maria 128
Kousis, Maria and Eder, Klaus 3, 26, 31, 92, 99, 101, 113, 120, 126, 138, 160
Kubicek, P. 17, 18, 149–51, 153

Latour, Bruno 120
Lewellen, Ted C. 13, 45, 66, 138, 148
local
  actions 128, 142
  campaigns 114, 130, 141, 168, 169
  collective action 135
  communities 128, 151
  concern 88, 141
  context 88, 169
  democracy 72
  ecosystem 125
  elections 83
  government 72
  group 136, 170
  opposition 81
  organisation 140
  people 154
  perception 88
  politics 63, 72
  reaction 135, 146
  social movements 137
  see also mobilisation, local; protest, local
localism 94

McAdam, Doug 132
McCartly, John 128
Mango, Andrew 16, 33–5, 39, 40, 156
Marcus, George 5–7, 11
Marx, Karl 131
Mayer, Margit 47, 48, 69, 72, 79, 81, 132
Melucci, Alberto 28, 131–3
methodological
  account 11
  approach 5
methodology 3, 29
Milton, Kay 31, 119, 122
mobilisation 1, 2, 11–14, 18, 21, 26, 28, 45, 46, 53, 61, 71, 72, 82, 84, 98, 101, 109, 111, 114, 117, 128, 132–5, 139, 142, 146, 150, 168
  local 82
movement 12, 14 , 15, 28, 46, 51, 78, 82, 90, 98, 108, 120–2, 128–44, 158, 160
  Bergama 111, 112
  civil 98
  culturalist 139
  democratic citizenship 111
  emancipation 142
  environmental 15, 99, 102, 103, 112, 114, 120, 122, 123, 132, 142, 144, 160, 167, 170
  feminist 141
  grass-roots 116, 142
  human rights 85, 142
  Islamist 144
  Kurdish 137
  minority 85
  New Social Movement 82, 131, 134, 135
  opposition 79, 139
  peace 132

peasant 12
political 128, 170
popular 133
protest 72
revolutionary 11, 131
social 2, 11–14, 19, 23, 28, 29, 44, 45, 48, 71, 72, 76, 77, 78, 81, 82, 92, 95, 98, 111, 115, 117, 119–22, 125, 128–44, 147, 157, 160, 166, 167, 169, 170
transcendentalist 122
transenvironmental 28, 117, 130
transnational 99, 142
urban 46, 48, 71, 72, 76, 170
Western Green 103
Wise Use 128
Young Turks, the 33

Nash, June 12, 14, 44, 47, 130, 134, 135, 137, 139, 141
Navaro-Yashin, Yael 12, 17, 36, 42, 97, 98, 115, 145, 146, 149, 153, 157, 159
New Social Movement Theory *see under* theory
neighbourhood
as social form 2, 158
NIABY 81
NIMBY 14, 76, 79, 81, 95, 108, 163
NIMBYism 80, 81, 133

Öncü, Ahmet and Koçan, Gürcan 1, 109–11, 128, 134, 137, 169
Özdemir, İbrahim 1, 18–20, 42, 100, 101, 103, 106, 109–12, 151, 155, 157, 160

participants (in ASG)
active 27, 36, 47, 49, 55, 60, 68, 77, 78, 81–4, 91, 94, 157, 165
less active 27, 55, 59, 60, 64, 77, 78, 84–9, 94, 126, 165, 166
non-active 27, 55, 59, 68, 78, 79, 88–90, 94, 166
supporters 27, 58, 78, 79, 90–3, 94, 117, 164, 165
protest 3, 4, 5, 10, 11, 12, 19, 22, 26, 27, 28, 29, 32, 45, 50, 51, 63, 67, 71, 74, 76, 79, 84, 86, 92, 102, 109, 110, 113, 115, 116, 124, 129, 134, 136, 140, 142, 163, 164, 166, 167, 168, 170
Akkuyu 140
anti-bridge 163
anti-dam 22, 114
ASG 73, 76, 164
Bergama 21, 109, 110, 111, 112, 114, 168
civil 1, 13, 75
environmental 74, 108, 128, 134, 135
grass-roots 1
local 2, 44, 114, 117, 125, 141, 169, 170
NIMBY 76, 79, 95, 108
Third Bridge 73, 76
transenvironmental 3, 26
urban 3, 27, 45, 55, 69, 72, 76, 94, 163
Psimitis, Michalis 2, 11, 14, 28, 131–4, 147

Resource Mobilisation Theory *see under* theory
Rollwagen, Jack 4, 13
Roots, Christopher 95, 99, 128, 129, 167

Sachs, Aaron 1, 110, 112, 128
Sassen, Saskia 5, 32, 45, 48
Shankland, David 12, 106
Şimşek, Sefa 15, 16, 132, 133, 137, 149–51
'Six Arrows' (*Altı Ok* – the six principles of Kemalism) 34
nationalism 34
populism 34
republicanism 34
revolution 34
secularism 34
as an identity 94, 145
and the army 154–7
statism 34

social movements *see under* movement, social
Social Movement Organisation 28
state 1, 9, 12, 13, 15, 16, 17, 18, 19, 21, 22, 33, 34, 43, 45, 47, 50, 51, 52, 53, 67, 72, 75, 76, 77, 85, 91, 93, 97, 98, 99, 100, 103, 105, 106, 107, 108, 111, 115, 116, 119, 121, 125, 128, 129, 132, 133, 134, 137, 138, 139, 140, 141, 143, 144, 145, 146, 147, 148, 149, 150, 151, 152, 153, 154, 155, 156, 157, 158, 159, 163, 164, 165, 167
  father 150
  nation-state 4, 32, 33, 97, 143, 163
  Turkish 3, 32, 39, 83, 84, 93, 97, 113, 164, 167, 168, 169
Stirling, Paul 33

Tarrow, Sidney 82
Theodossopoulos, Dimitrios 81, 135, 136
theory
  Dependency Theory 44
  Global Social Movement Theory 14, 134
  New Social Movement Theory 12–14, 131–5
  Resource Mobilisation Theory 12, 13, 132–4
  World System Theory 5, 137
Tilly, Charles 132
transenvironmental
  campaign 17, 21, 29, 97, 99, 102, 114, 115, 117, 168
  character 55, 167
  collective action 130
  group 93
  mobilisation 55
  protest 3, 26
  social movement 23, 167, 170
  see also movement, transenvironmental
transenvironmentalism 55, 163, 166
transnational 1, 5, 13, 15, 44, 45, 47, 48, 50, 53, 55, 85, 86, 87, 99, 110, 111, 120, 136, 137, 138, 139, 141, 142, 143, 144, 160, 169, 170
transnationally 93, 111, 160
transnationalism 140
Turkey 1, 3, 5, 6, 7, 8, 9, 15, 16, 17, 18, 19, 20, 21, 24, 29, 31, 32, 34, 35, 37, 38, 39, 40, 41, 43, 45, 46, 47, 49, 50, 51, 53, 67, 70, 75, 77, 82, 83, 84, 85, 86, 87, 91, 93, 97, 98, 99, 100, 101, 102, 103, 104, 105, 106, 107, 108, 110, 112, 113, 116, 120, 123, 130, 140, 142, 144, 145, 146, 149, 150, 151, 152, 153, 154, 155, 156, 157, 158, 159, 160, 163, 164, 167, 169; *see also* state, Turkish

urban
  action 3
  character of ASG 27
  conflict 32, 48, 93
  demand 72, 93
  environment 20, 31, 100
  face of ASG 27
  milieu 31
  policy 32
  setting 44
  social movements 45, 46, 48, 72
  studies 1, 45
  systems 4
  see also anthropology, urban; movement, urban; protest, urban

Van Naerssen, Ton 69, 72

Wallerstein, Immanuel 5, 44, 46
Wapner, Paul Kevin 129, 138, 143, 147
Wolsink, Maarten 14, 81

Zald, Mayer 132

FSC
www.fsc.org
MIX
Papier | Fördert gute Waldnutzung
FSC® C083411

Kolibri 360 GmbH
Ferdinand-Jühlke-Straße 7
99095 Erfurt, Deutschland
produktsicherheit@kolibri360.de